ANALYTICAL MODELLING OF STRUCTURAL SYSTEMS
An Entirely New Approach with Emphasis on Behaviour of Building Structures

ELLIS HORWOOD SERIES IN CIVIL ENGINEERING

Series Editors
Structures: Professor H. R. EVANS, Department of Civil Engineering, University College, Cardiff
Hydraulic Engineering and Hydrology: Dr R. H. J. SELLIN, Department of Civil Engineering, University of Bristol
Geotechnics: Professor D. M. WOOD, Department of Civil Engineering, University of Glasgow

ANALYTICAL MODELLING OF STRUCTURAL SYSTEMS

An Entirely New Approach with Emphasis on Behaviour of Building Structures

IAIN A. MACLEOD B.Sc., Ph.D., C.Eng., F.I.Struct.E., F.I.C.E.
Professor of Structural Engineering
Department of Civil Engineering
University of Strathclyde

ELLIS HORWOOD
NEW YORK   LONDON   TORONTO   SYDNEY   TOKYO   SINGAPORE

First published in 1990 by
ELLIS HORWOOD LIMITED
Market Cross House, Cooper Street,
Chichester, West Sussex, PO19 1EB, England
A division of
Simon & Schuster International Group

© Ellis Horwood Limited, 1990

Typeset in Times by Ellis Horwood
Printed and bound in Great Britain
by The Camelot Press, Southampton

British Library Cataloguing in Publication Data

Macleod, Iain A.
Analytical modelling of structural systems: An entirely new
approach with emphasis on behaviour of building structures.
1. Structural engineering. Mathematics. Finite element
methods
I. Title
624.1′01′515353
ISBN 0–13–035254–3 (Library Ed.)
ISBN 0–13–035247–0 (Student Pkb. Ed.)

Library of Congress Cataloging-in-Publication Data

MacLeod, Iain A. (Iain Alasdair), 1939–
Analytical modelling of structural systems: An entirely new
approach with emphasis on behaviour of building structures/
Iain A. MacLeod.
p. cm. — (Ellis Horwood series in civil engineering)
ISBN 0–13–035254–3 (Library Ed.)
ISBN 0–13–035247–0 (Student Pkb. Ed.)
1. Structural analysis (Engineering)—Data processing
I. Title. II. Series.
TA647.M33 1990
624.1′71—dc20   89–29860
                                             CIP

Table of contents

To Barbara, Mairi and Alastair

Acknowledgements

My introduction to structural analysis was from Hugh Nelson and the late Bill Marshall at the University of Glasgow and I owe a lot to their early guidance. I have received much support from teaching and research colleagues and from research students. They include P. Arthur, P. Bhatt, D. Green, G. Pole, I. A. Smith, I. M. Smith, H. B. Sutherland, W. Wilson, A. Derecho, M. Fintel, C. Freyermuth, A. Gouwens, M. Haddadin, P. Mast, W. Bannatyne, A. Clark, J. Duncan, R. Hardie, D. MacKenzie, S. Abu el Magd, R. Gibson, A. Gillespie, H. Hosny, R. Mills, L. Smith, M. Zacas, J. Allan, A. Buchan, R. Cairns, D. Clarke, D. Logan, J. Marshall, A. Ritchie, N. Ades, A. Houmsi, F. Kor, T. Sreetheran, M. Suppiah.

I have also much appreciated the help given to me by many friends in engineering practice.

I would like to thank the governors of Paisley College of Technology for granting me six months study leave in 1980 to work on the book. The support of the late Tom Howie, then Principal of the College was much appreciated. The study leave was spent in Australia and I record my appreciation to Douglas Clyde of the University of Western Australia and Ian Lee then of the University of New South Wales and their respective colleagues for the kind reception and assistance that they gave me at that time.

In the direct preparation of the book my thanks go to:

— Bill Duncan, Mohamed Rafiq, Akin Ogundiya and Muslikh who helped with examples and reading drafts,
— Sheena Nelson, Elanor Caldwell and Una White who did the diagrams,
— Rosaleen Lynch and my wife Barbara who typed the manuscript.

Finally in creating the time to write the book my family had to be neglected to a degree and I would like to express my thanks to them for their love and support.

1

Introduction

1.1 GENERAL

This book is about how to set up and check computer-based models of structural systems. It is for users of software rather than for programmers. Its format is based on the principle that knowledge and processing are best separated where possible. Thus Chapters 1 to 7 concentrate mainly on knowledge and Chapter 8 concentrates on processing.

Chapter 4 represents the heart of the knowledge. This chapter gives advice on how to formulate analytical models and readers may wish to go directly to it since some the discussion on material and element behaviour given in Chapters 2 and 3 may seem elementary. However, many people who use structural analysis packages are fairly ignorant about basic assumptions, knowledge of which is essential to the reflective approach advocated in this text.

Chapter 5 and 6 give advice on how one can check computer results and Chapter 7 discusses modelling of dynamic situations.

In Chapter 8, definitions, mathematical derivations and solution processes are discussed.

The separation of knowledge and processing

Kowalski (1979) uses the relationship

$$A = L + C$$

where A represents an algorithm, L is the logic, or knowledge component, C is the control, or processing part.

In programs, the L and C components are mixed together, thus obscuring them both. However, if the knowledge (in the form of facts and rules) is stored separately in a 'knowledge base' and operated on by a processor then both components can be more readily understood. This is an extremely useful concept in computer-aided design. Designers can concentrate on knowledge and spend less time on the details of processing.

This de-emphasis of processing breaks a long tradition in structural analysis. There has been a widely shared view that an understanding of structural behaviour springs mainly from doing hand calculations. While it is important that engineers can do calculations and carry out mathematical manipulations, the process of doing such tasks seldom adds significantly to understanding in my experience. The most important aspect of processing is that it can be shown to have been done correctly. Details of how it is done need concern us less.

In a computer-based environment, the most important factors in analytical modelling are

- understanding of basic assumptions
- understanding of structural behaviour
- ability to create and validate analytical models.

That understanding the means of achieving a solution is now of lesser importance is viewed by some as being heretical. Traditionalists view the use of a computer to produce solutions as being neutral or possibly negative to understanding of structural behaviour. Those who do use computers know that this is not so. The computer frees us from the shackles of processing and can promote a depth of understandilng that traditional techniques cannot deliver.

A reflective approach

The transformation of computer use from a neutral to a positive agent for under-standing depends on the adoption of a reflective attitude. Such an attitude involves results always being viewed with suspicion and never accepted without question. The process of looking at results in this way can be rewarding. Insights into behaviour are often gained, leading to enhancement of the knowledge of the user.

If a reflective approach is not used then the computer begins to control the design process, which is manifestly unaceptable.

Validity of assumptions

Modern scientific opinion (Popper 1977) suggests that a theory cannot be proved to be correct in an absolute sense. It can only be shown to be not false to some degree. Good theories are strongly not falsified; poor theories are weakly not falsified. Thus there are no basic assumptions that can be described as 'exact'. In other words, the validity of an assumption is *always* a relative assessment.

In this text, when discussing assumptions, the description 'good' means that the assumption (under the stated conditions) represents real behaviour fairly accurately. The description 'acceptable' means that the assumption may be rather rough but can nevertheless give useful results. 'Valid' is either good or acceptable.

1.2   BASIC PRINCIPLES IN MODELLING

(1) Never use a model that is more complex than is necessary. The simplest models are often the most useful.

(2) Treat results in the opposite way to an accused person in law. Results should be considered to be guilty (of being in error) until they have passed rigorous scrutiny.

(3) If an effect is included in the analytical model then it must be taken into account when sizing the member otherwise the lower bound theorem will not be valid. For example, if bending deformation is included in the analysis of a truss, the members must be sized for axial and bending actions.

(4) Just because all checks point to the conclusion that the results are valid, does not mean that they are indeed correct.

(5) Do not attempt to model detailed behaviour as part of a large model. Extract the part of the large model that needs detailed treatment and create a special model for it. See, for example, the model illustrated in Fig. 4.11.

1.3   A HIERARCHY OF MODELS

The purpose of modelling is to try to represent the behaviour of a real (or potentially real) object. We call this object the 'prototype'.

The real behaviour of the prototype is normally highly complex and many simplifications have to be made in order to model it. Material behaviour has to be simulated; the means of support and the ways that the parts fit together need to be modelled; the loading needs to be defined.

There are three basic steps in the modern approach to creating an analytical model of a structure:

(1) Assumptions are made about the behaviour of a small piece of the material — denoted here as a 'differential element'. This forms the 'material model'. For structural problems the main assumptions are the relationships between stress and strain — normally called the 'constitutive' relationship.

(2) An integration process is performed on the differential element to define the behaviour of an element which represents a finite part of the structure — hence the term 'finite element'.

(3) The final level is an assembly of finite elements which represents the behaviour of the complete structure — the finite element model. At this stage the boundary restraints are imposed and loading conditions defined.

Some problems can be solved without using the finite element concept. This is achieved by defining the behaviour as a continuous function over the structure and creating a solution for given boundary conditions and loading. Such solutions are often referred to as being 'exact'. This term is used here to describe solutions of this type but it must be remembered that they are only exact in relation to the assumptions made and may not relate at all to prototype behaviour. Exact solutions are only feasible for a limited class of structural problem and modern computing power makes the use of finite elements the most versatile and effective approach.

Four levels in the modelling hierarchy when using finite elements can therefore be defined:

(A) The real behaviour of the prototype.
(B) The behaviour of the differential element.
(C) The behaviour of a single finite element.
(D) The behaviour of the finite element model.

In a given situation we need to assess the relationship of (A), (B) and (C) with (D)

Important assumptions are made at all levels and one should maintain a sense of proportion when choosing elements. Although the use of finite elements does constitute an extra level of assumption, the major assumptions may lie elsewhere. The constitutive relationship is often a very crude approximation and boundary conditions and loading can be dominant factors.

Researchers often make the mistake of tuning the choice of elements to achieve correlation with protoype measurements when there may be other features of their model which are causing discrepancies.

1.4   STRUCTURE OF THE TEXT

This text has the following main objectives:

- To present basic essential knowledge.
- To promote a reflective approach to the activities involved in modelling.
- To give advice about how to establish and check analytical models.

Readers are expected to have a basic knowledge of structural mechanics, matrix notation and the stiffness method.

Matrix notation is used where it is relevant. For those who are not accustomed to its use, the time spent in becoming familiar with it will be well rewarded. It allows the processing steps to be expressed in a succinct way that, for me, greatly clarifies their meaning. I cannot imagine how the concepts used in modern analytical modelling could have developed without matrix notation.

The examples given relate mainly to building structures but the principles of modelling which are discussed in the text relate to all types of structure.

The text is not intended to be read sequentially. Mathematical processing steps are mainly omitted from the early chapters so as to concentrate on the assumptions but readers may find it worthwhile to move quickly to Chapter 8 which gives derivations.

The text is structured in relation to the steps in modelling rather than in relation to types of problem. These steps are

- Decide on the material model — Chapter 2.
- Decide on the elements to be used — Chapter 3.
- Form the model of the structure — Chapter 4.
- Check the results — Chapters 5 and 6.

1.5   TECHNIQUES FOR UNDERSTANDING BEHAVIOUR
1.5.1   Physical observation and testing

All analytical models are based on physical observations and the importance of this for understanding behaviour cannot be understated.

Structural behaviour is not easily observed. Structures normally move only imperceptibly and casual observations may yield little information.

A collapse is a major source of knowledge although such events are rare nowadays. When a collapse does occur one should try to extract the maximum amount of information from it.

Some people have an instinctive feeling for structural behaviour and knowledge can be gained from observing the products of such ability. Good structures have a logicality to them that can normally be explained and critical analysis of such systems can help the less gifted to improve their own performance.

Prototype observations

The real *in situ* behaviour of a structure is the ultimate test of validity and much useful information is gained from such observations. There are, however, major difficulties in doing this. These include:

- The applied conditions (load, temperature, etc.) are difficult to assess.
- In buildings, 'non-structural' components such as partitions and cladding affect the behaviour.
- Analytical models of building structures tend not to predict real behaviour.

Real events can sometimes make fundamental changes to our knowledge of behaviour. A good example of this is in the design of earthquake-resistant buildings. Prior to observation of collapsed buildings such as in Anchorage, Alaska, and Caracas, Venezuela, the conventional approach for earthquake design was to provide flexible moment resisting frames. The philosophy was that such frames had a lower natural period; this would cause lower accelerations in the structure and, hence, lower forces. More rigid structures were felt to be more susceptible to damage. The observations showed that the opposite was the case. Rigid buildings with properly designed shear walls perform much better than those with moment resisting frames and our approach to earthquake design has been radically altered by the observations of real performance.

Prototype testing

Controlled *in situ* tests can provide useful information but normally only for localized features, e.g. a slab or beam test or a fixing test.

Prototype tests tend to validate only certain aspects of analytical modelling. For example, a load test on a bridge shows that the use of a model has resulted in a structure that meets the design requirements. This does not necessarily mean that the model provides a good representation of the structure unless monitoring of stress and deformation has been carried out.

Laboratory testing

Here the applied conditions can be carefully controlled; again only components of a real system can be included.

1.5.2 Understanding basic assumptions

The basic assumptions used in analytical models (Chapters 2 and 3) are themselves models of the behaviour and one should not attempt structural analysis without a good appreciation of them.

1.5.3 Understanding the mathematical basis of models

The mathematics is part of the model and whether it aids understanding depends on:

- the complexity of the mathematics.
- the mathematical ability of the user.

A large proportion of civil engineers are somewhat uncomfortable with mathematics despite having studied it for upwards of seven years. This may be due to the normal method of presentation at school and at university which tends to disregard applications. Mathematics should be treated as a useful tool rather than something to be avoided if possible. Basic mathematical statements such as governing equations and boundary conditions are concise statements of assumptions. The processing of these equations to provide solutions performs a different function and the depth of understanding needed by the user for such activity is subject to debate. One should try to develop a general understanding of the mathematical processing that is being used but in many cases a deep understanding may not be necessary (nor possible).

1.5.4   Parameter studies

With new cheap processing power, several runs on a model with variations in one aspect can normally be made at little extra cost. The effect of these variations can then be studied and conclusions drawn. Typical results of such activity are:

- Identification of the sensitivity of design parameter values to assumptions made. For example, in the analysis of a moment resisting frame, one can vary the member stiffnesses to assess whether or not this affects the design moments significantly.
- Insight into behaviour. An example of this is given in section 5.3.3.

1.5.5   Use of simplified models

A simplified model is one that can be done by hand, or at least, without extensive computing. If these are used in conjunction with more sophisticated models, important insights may be gained. Such models are discussed in Chapter 6.

1.6   EXPECTATION OF ACCURACY

In a stiffness analysis one obtains, firstly, the deformations; these are used to obtain member forces and stresses. To do this, operations on the deformations are performed which are equivalent to differentiating them. Numerical differentiation gives results which are less accurate than the values on which the differentiation is performed (and, conversely, integration tends to improve accuracy). Therefore one tends to find that deformations are predicted to better accuracy than internal forces and stresses. This is so when one compares the results with more accurate solutions based on the same material behaviour.

However, in an elastic analysis, the internal actions are not dependent on the absolute values of stiffness. They depend on the relative stiffnesses and if the E value is constant, it does not affect the distribution or magnitude of the stresses. These stresses are in equilibrium with the applied load and if used to proportion the member sizes will tend to give safe results (using the lower bound theorem (section 8.3.1)). Therefore although the prediction of absolute deformation could be in error to an unsatisfactory degree the corresponding internal actions can be used to good effect.

In Section 1.5.4 the use of parameter studies is discussed. When doing such studies we assume that the relative effects are predicted more accurately than the absolute behaviour. The philosophy of using such a principle is discussed in section 4.5.1.

Our analytical models, imperfect as they are, are often the best we can do in the circumstances. They can provide useful, if incomplete, information. The reflective designer does not treat the results of an analysis with greater confidence than is merited but uses them as a useful tool to help to design good structures.

2

Material behaviour

2.1 GENERAL

The initial step in creating an analytical model is to define the behaviour of the material. In this chapter the assumptions made and the commonly used constitutive relationships are discussed.

Attributes of a constitutive relationship are:

- It relates stress σ to strain ε.
- A generalized definition of 'stress' and 'strain' is used such that a constitutive relationship can relate force to displacement or moment to rotation, etc.
- It relates to a differential element which is small enough such that the stresses are uniform over the area or over the lengths of the element.

Although some of the relationships discussed in this chapter might be considered elementary they are included to focus attention on their importance. They represent the basic level of modelling which, if invalid, negate the whole concept of the model being used.

Our ability to predict real behaviour is now much more limited by lack of good descriptions of behaviour than by processing difficulties. Often an elastic solution will give just as much (or as little) useful information as a complex non-linear analysis if the latter model ignores important features of behaviour. For example, an elastic solution gives no information about the long term deformation of a concrete beam but a solid element model using non-linear material properties might not give better results unless cracking, creep and shrinkage are properly modelled.

Derivations for constitutive relationships are given in section 8.7.

Notation

σ_i — direct stress in direction i
τ_{ij} — shear stress on i plane in the j direction
E_i — Young's modulus in direction i
v — Poisson's ratio

ε_i — Direct strain in direction i
γ_{ij} — Shear strain on the i plane in the j direction
P — Axial load
A — Cross-sectional area
x,y,z — Cartesian coordinate directions
u,v,w — deformations in the x,y,z directions respectively
Symbols printed in bold represent matrices
ε — strain vector
σ — stress vector
$\mathbf{D}$ — constitutive matrix.

2.1.1   General form of the constitutive relationship

Before any load is applied, the material may be subject to 'initial strains' ε_0 and 'initial stresses' σ_0. After application of the load, the strains are ε and the stresses are σ. The constitutive relationship then has the form:

$$\sigma = \mathbf{D}(\varepsilon - \varepsilon_0) + \sigma_0 \qquad (2.1)$$

where $\mathbf{D}$ is the constitutive matrix.

The initial strains and stresses tend to be used only in non-linear analysis and the following shortened form of equation (2.1) is normally quoted in this text:

$$\sigma = \mathbf{D}\varepsilon \qquad (2.2)$$

2.1.2   Small deformations

When an element of material is subject to an external action (e.g. load variation, temperature variation, etc.) its shape changes. It is assumed that this change in shape has only a second order effect on the behaviour. For example, if a direct force P is applied over an area A, the area will change by an amount δA. The direct stress (σ) is defined in terms of the original area, i.e.

$$\sigma = P/A \text{ and } not \quad \sigma = P/(A + \delta A)$$

This is a basic assumption which is seldom unrealistic in structural engineering situations. However, moment caused by the eccentricity of axial load due to lateral deformation can be significant — see section 2.4.

2.1.3   Linear elasticity

The term 'elastic' means that deformation is uniquely defined by the load, i.e. at a given load level the deformation state is the same whether the material is being loaded or unloaded.

Linear elasticity is the most common material model in stuctural engineering. It is based on the following assumptions:

- As stress increases the resulting strain increases in a linear proportion.
- As stress decreases the resulting strain decreases in the same linear proportion.
- Strain induced at right angles to an applied strain is linearly proportional to the applied strain (Poisson's ratio effect).
- The material is homogeneous and continuous.

At low stress levels the linear stress–strain assumption is valid for a wide range of engineering materials. For example:

- For metals at normal temperatures the assumption is good.
- For concrete and masonry without tensile cracking and under short term loading it is less accurate but acceptable.
- Timber is not isotropic but the elastic model is good in uniaxial situations.
- The linear model for soil is acceptable only for short term deformation of stiff clays and sands. It is used for soils as a means of assessing the effect of foundation movement on the superstructure as described in Section 4.5.

Whether or not a material can be described as continuous depends of the viewpoint. Masonry does not have the appearance of a continuous material but steel does. However, if we examine the microstructure of steel we find that there is a high degree of discontinuity. Therefore a large piece of masonry may be no less discontinuous than a small piece of steel.

2.2   LINEAR RELATIONSHIPS

In this section some commonly used linear constitutive relationships are discussed.

2.2.1   Plane stress

The isotropic plane stress relationship is:

$$\begin{Bmatrix} \sigma_x \\ \sigma_y \\ \tau_{xy} \end{Bmatrix} = \frac{E}{1-v^2} \begin{bmatrix} 1 & v & 0 \\ v & 1 & 0 \\ 0 & 0 & \dfrac{1-v}{2} \end{bmatrix} \begin{Bmatrix} \varepsilon_x \\ \varepsilon_y \\ \gamma_{xy} \end{Bmatrix} \qquad (2.3)$$

The stress directions are shown in Fig. 2.1. Note that the first subscript in the shear stress symbol refers to the plane in which the stress acts. The second subscript refers to the direction of the stress. For example, τ_{xy} acts on the x plane (which is at right angles to the x-direction) in the y-direction.

Plane stress is based on the following conditions:

- there is no applied stress in the z-direction, i.e. $\sigma_z = 0$.
- there is no restraint to strain in the z-direction, i.e. ε_z unrestrained.
- values of E and v are established from tensile test measurements.

This relationship is derived in section 8.7.1.

2.2.2   Plane strain

When there is restraint in the z-direction, for example when considering slices of long uniform structures such as dams and embankments, then it is normal to assume full restraint in the z-direction, i.e. $\varepsilon_z = 0$. This gives the plane strain condition which for isotropic conditions is defined by:

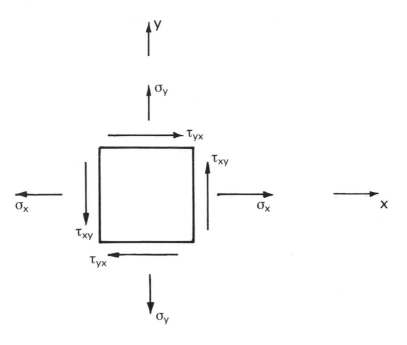

Fig. 2.1 — Stresses for plane stress and plane strain.

$$\left\{\begin{array}{c} \sigma_x \\ \\ \sigma_y \\ \\ \sigma_{xy} \end{array}\right\} = \frac{E(1-v)}{(1+v)(1-2v)} \begin{bmatrix} 1 & \dfrac{v}{1-v} & 0 \\ \dfrac{v}{1-v} & 1 & 0 \\ 0 & 0 & \dfrac{(1-2v)}{2(1-v)} \end{bmatrix} \left\{\begin{array}{c} \varepsilon_x \\ \\ \varepsilon_y \\ \\ \gamma_{xy} \end{array}\right\} \qquad (2.4)$$

σ_z is not zero and can be calculated from

$$\sigma_z = v(\sigma_x + \sigma_y) \qquad (2.5)$$

Note that if plane strain is not directly available then the plane stress conditions can be substituted with values of E' and v' as follows:

use $E' = E/(1 - v^2)$
use $v' = v/(1 - v)$

The plane strain relationship is derived in section 8.7.2.

2.2.3   Isotropic 3D
For a differential element subject to stress in three dimensions (Fig. 2.2) the constitutive relationship is

$$\begin{Bmatrix} \sigma_x \\ \sigma_y \\ \sigma_z \\ \tau_{xy} \\ \tau_{xz} \\ \tau_{yz} \end{Bmatrix} = \frac{E}{(1+v)(1-2v)} \begin{bmatrix} 1-v & v & v & 0 & 0 & 0 \\ v & 1-v & v & 0 & 0 & 0 \\ v & v & 1-v & 0 & 0 & 0 \\ 0 & 0 & 0 & \frac{1-2v}{2} & 0 & 0 \\ 0 & 0 & 0 & 0 & \frac{1-2v}{2} & 0 \\ 0 & 0 & 0 & 0 & 0 & \frac{1-2v}{2} \end{bmatrix} \begin{Bmatrix} \varepsilon_x \\ \varepsilon_y \\ \varepsilon_z \\ \gamma_{xy} \\ \gamma_{xz} \\ \gamma_{yz} \end{Bmatrix} \qquad (2.6)$$

This relationship is derived in section 8.7.4.

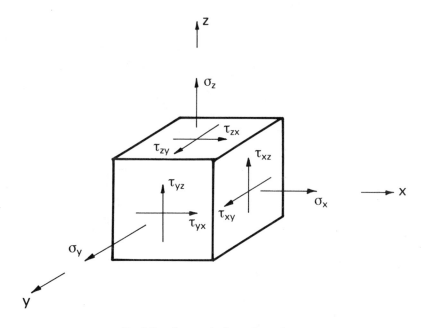

Fig. 2.2 — Stresses in three dimensions.

2.2.4 Beam bending

The constitutive relationship for beam bending is:

$$M = EI\ \mathrm{d}^2v/\mathrm{d}x^2 \qquad (2.7)$$

where M is the bending moment
I is the second moment of area of the section
(Table A1 gives formulae for I values)
v is the displacement in the y-direction

See Fig. 2.3.

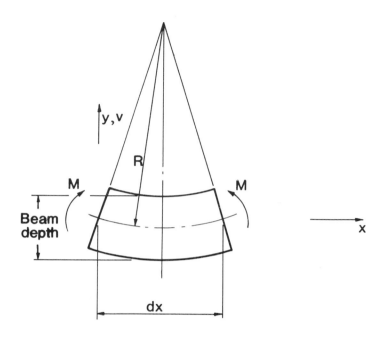

Fig. 2.3 — Bending differential element.

The assumptions for beam bending in addition to elasticity are:

(1) Plane sections remain plane. For prediction of direct bending stress and shear stress this assumption gives results as follows:
   - with high span to depth depth ratios — good.
   - with deeper beams — span to depth ratios in range 3.0 to 6.0 — acceptable
   - with low span to depth ratios (less than 3.0) — accuracy may be low but order of magnitude estimates may be obtained in some cases.

(2) Direct stress at right angles to axis of beam. This is neglected, giving good results except in the vicinity of concentrated loads.

(3) Small deformations. The use of this assumption in the development of equation (2.7) is acceptable at the differential element level but second order effects may need to be considered for bending elements as discussed in section 3.11.

(4) Shear deformation can be calculated independently of bending deformation. This results in good accuracy for estimates of deformation except with very low span-to-depth ratios.

2.2.5   Thin plate bending

Elastic thin plate bending is one of the more commonly used material models in structural engineering. Fig. 2.4 shows the forces acting on a thin plate element. The notation and sign convention used follows that of Timoshenko and Woinowsky-Krieger (1959). Note that the direct moments M_x and M_y and the twisting moment M_{xy} are moments *per unit width* of the plate.

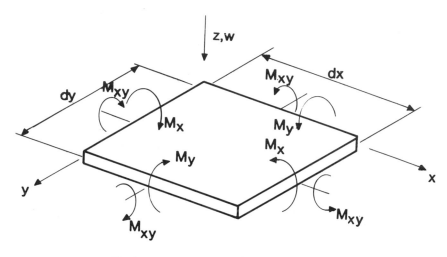

Fig. 2.4 — Internal moments on a plate element.

Isotropic thin plate

The constitutive relationship for isotropic conditions is as follows:

$$\begin{Bmatrix} M_x \\ M_y \\ M_{xy} \end{Bmatrix} = D \begin{bmatrix} 1 & v & 0 \\ v & 1 & 0 \\ 0 & 0 & \dfrac{1-v}{2} \end{bmatrix} \begin{Bmatrix} -\dfrac{\partial^2 w}{\partial x^2} \\ -\dfrac{\partial^2 w}{\partial y^2} \\ 2\dfrac{\partial^2 w}{\partial x \partial y} \end{Bmatrix} \qquad (2.8)$$

where $D = \dfrac{Eh^3}{12(1-v^2)}$

    w is the displacement of the plate in the z-direction,
    M_x, M_y are the direct moments per unit width — Fig. 2.4,
    M_{xy} is the twisting moment per unit width — Fig. 2.4.

This relationship is developed in section 8.7.5.
The twisting deformation is a derivative of shear strain.

$$\text{shear strain} = \frac{\partial}{\partial x}\frac{\partial w}{\partial y} + \frac{\partial}{\partial y}\frac{\partial w}{\partial x} = 2\frac{\partial^2 w}{\partial x \partial y}$$

Thin plate bending is based on the following assumptions:

- Linear elastic behaviour — see section 2.1.3.

- Normals to the plane of the plate remain straight after deformation. This assumption is satisfactory while shear deformation can be neglected.
- Shear deformation is neglected. This is valid for thin plates. With span-to-depth ratios greater than 10 : 1 there is no need to consider shear deformation and useful results can be obtained with lower span to depth ratios.
- The deflections are small in comparison with the plate dimensions. A reasonable limit on this is that the maximum deflection should not be greater than the plate depth. Such a criterion is seldom violated in conventional structures. The second order effect of axial loads may need to be taken into account in some cases — see section 4.8.4.
- There is no strain in the centroidal plane of the plate. This assumption can be easily removed by adding in-plane actions to the bending actions as used for flat shell elements — section 3.7.

Orthotropic thin plate
The orthotropic thin plate relationship is normally quoted as:

$$\begin{Bmatrix} M_x \\ M_y \\ M_{xy} \end{Bmatrix} = \begin{bmatrix} D_x & D_1 & 0 \\ D_1 & D_y & 0 \\ 0 & 0 & D_{xy} \end{bmatrix} \begin{Bmatrix} -\dfrac{\partial^2 w}{\partial x^2} \\ -\dfrac{\partial^2 w}{\partial y^2} \\ 2\dfrac{\partial^2 w}{\partial x \partial y} \end{Bmatrix} \tag{2.9}$$

Assignment of values to the constants of equation (2.9) are discussed in section 4.4.4.

2.2.6 Shear deformation in bending
The constitutive relationship for shear deformation in bending is:

$$S = \overline{A} G \, dv/dx \tag{2.10}$$

where S is the shear force, $\overline{A}$ is the shear area — Table A2, dv/dx is the rotation due to shear — see Fig. 2.5.

Fig. 2.5 shows a differential beam element under shear deformation only. The shear stress distribution over the cross-section is parabolic resulting from the linear direct bending stress. There is no shear strain at the top and bottom of the cross-section and the angle of the cross-section to the longitudinal line of the beam remains at 90° at these levels. The deformed cross-section of the beam is therefore in double curvature as shown.

Equation (2.10) is based on two main assumptions:

- The shear stress distribution results from the linear bending stress distribution.
- the cross section is free to warp i.e. plane sections do not remain plane.

These assumptions in combination do not appear to be promising since they seem to compete with each other. However, since the bending and shear components are treated as being uncoupled, predictions of beam deformation resulting from the addition of bending and shear components can give good results in many cases — see Case Study 3.1 (p. 42).

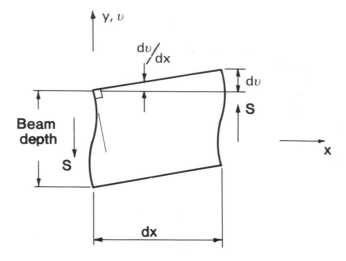

Fig. 2.5 — Shear deformation of a differential beam element.

2.2.7 St Venant torsion

The constitutive relationship for St Venant torsion is

$$dT/dx = CG \; d^2\varphi/dx^2 \tag{2.11}$$

where dT/dx is the torque per unit length, C is the torsion constant (Table A3), G is the rigidity modulus, φ is the angle of twist of the cross-section (Fig. 2.6(a)).

This is based on the following assumptions:

- with constant torque a straight line on the element remains straight after the torque is applied (Fig. 2.6(a)).
- The cross section is free to warp.

St Venant torsion can give good results for closed sections. For open sections warping torsion needs to be included in the model.

2.2.8 Warping torsion of thin-walled sections

This form of torsion depends on warping restraint and is applicable to sections made up from flat plates (thin-walled sections) — Fig. 2.6(b). The constitutive relationship for warping torsion is:

$$dT/dx = EI_\omega \; d^4\varphi/dx^4 \tag{2.12}$$

where EI_ω is the warping stiffness, I_ω is the second moment of 'sectorial' area.

The main assumptions are;

- the plates which form the cross-section deform in bending in their own planes,
- out-of-plane bending of the plates is neglected,
- shear deformation is neglected,
- the plates are continuously connected to each other longitudinally.

The degree of warping restraint may not be easy to define and this may cause the

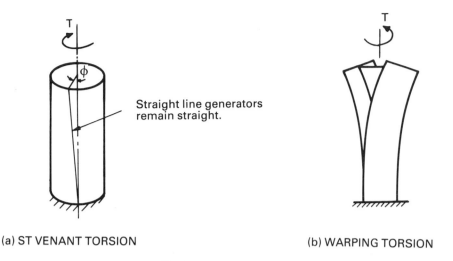

(a) ST VENANT TORSION   (b) WARPING TORSION

Fig. 2.6 — Torsion models.

warping torsion to be less accurate. For most open sections, warping and St Venant torsion need to be combined — see section 2.2.9.

Warping torsion is not well understood by most structural engineers mainly because structural members are not often required to resist torsional loading; when they do a closed section is normally used. Another factor may be the relative mathematical complexity of warping torsion theory. Vlasov's theory for warping torsion (Vlasov 1961, Zbirohowski-Koscia 1967) is not difficult to use once the basic concepts are understood but the theory is seldom described in texts on structural mechanics.

2.2.9 Combined torsion

The constitutive relationship for combined St Venant and warping torsion is:

$$dT/dx = EI_\omega \, d^4\varphi/dx^4 - CG \, d^2\varphi/dx^2 \qquad (2.13)$$

The torque T is considered to be resisted by two components

T_ω the flexural twisting moment and
T_v the St Venant twisting moment
i.e. $T = T_\omega + T_v$ $\qquad (2.14)$

The bimoment B is defined as

$$B = \int T_\omega dx \qquad (2.15)$$

This is the warping torsion equivalent of a bending moment. In Vlasov's theory, for example, the direct stress due to warping torsion is

$$\sigma = \frac{B\omega}{I_\omega} \qquad (2.16)$$

where ω is the sectorial coordinate of the point at which stress is required. The equivalence of equation (2.16) and that for direct stress due to bending $\sigma = My/I$ is readily identified.

The combination of St Venant and warping torsion gives a model for torsion which can give reasonable predictions of behaviour. The accuracy in a given situation is, however, not easy to predict.

2.3 NON-LINEAR MATERIAL BEHAVIOUR

It is common to want to investigate design situations where the material behaviour is not linear elastic. Typical situations are ultimate load analysis, long term deformation of concrete and soils and earthquake analysis.

'Plasticity' is a condition in which non-recoverable deformation can take place. Most structural materials behave elastically up to a limit and then fail (brittle material) or behave plastically.

The most common material model for plasticity is 'perfectly plastic', i.e. where beyond the plastic limit it continues to withstand its yield stress — σ_y — but does not take further stress (see Fig. 2.7). Mild steel behaves fairly closely to this ideal. For most other ductile materials it is a fairly rough approximation.

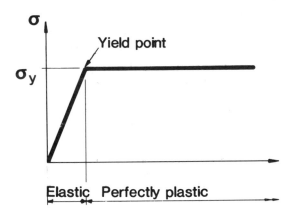

Fig. 2.7 — Elastic — perfectly plastic stress-strain diagram.

A most important factor in relation to plasticity is ductility. This is the deformation capacity beyond yield normally defined by a ductility factor μ where

$$\mu = \frac{\text{strain at ultimate load}}{\text{strain at yield load}}$$

or

$$\mu = \frac{\text{deformation at ultimate load}}{\text{deformation at yield load}} \qquad (2.17)$$

When using a plasticity assumption, the fact that no structural material has unlimited ductility must be taken into account. This is further discussed in section 8.3.

Another common non-linear material relationship is elastic–plastic with strain hardening — see Fig. 2.8. If unloading takes place from beyond the yield level — σ_y — it is assumed that the stress–strain relationship during unloading is linear and parallel to the elastic loading curve — see Fig. 2.8. At zero load after unloading, there will be a residual plastic strain ε_p.

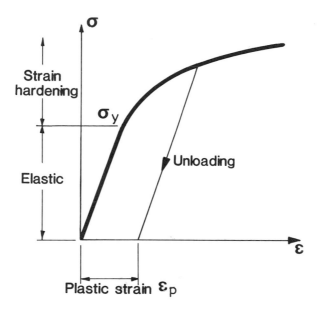

Fig. 2.8 — Elastic — strain hardening stress-strain diagram.

'Yield' is the point at which some of the deformation becomes non-recoverable. With uniaxial stress (e.g. in the tensile test) the decision as to the point of yield may not be easy to define precisely since there is not normally a sharp transition between elastic and plastic zones. However, a yield stress is defined as σ_y, marking the onset of plasticity.

With biaxial or triaxial stress the onset of plastic behaviour is less easy to define and a number of yield criteria have been developed for such situations. For metals, the Tresca and the Von Mises yield criteria appear to give good results in many cases. The Tresca condition is based on maximum shear stress and is more conservative than Von Mises which is based on maximum shear strain energy. For soils and concrete the Mohr–Coulomb and Drucker–Prager criteria are more relevant.

Associated with plastic behaviour is a 'flow rule' which defines the post-yield deformation characteristics.

Modelling of non-linear situations is a developing and important subject. Virtually any stress–strain relationship can now be handled computationally but specialized knowledge is needed on the part of the user.

Non-linear behaviour of concrete structures is an especially important topic (Branson 1977, Gilbert 1988).

2.4   NON-LINEAR GEOMETRY

A strut which is straight when unloaded takes an axial applied load N and is given a small lateral displacement v — Fig. 2.9(a). Moment equilibrium of a length of the strut — see Fig. 2.9(b) — gives:

$$Nv + M = 0 \tag{2.18}$$

M is taken as sagging positive, hence

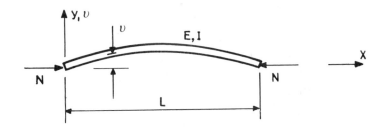

(a) STRUT WITH LATERAL DISPLACEMENT

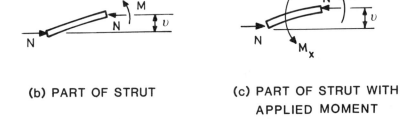

(b) PART OF STRUT   **(c) PART OF STRUT WITH**
                               APPLIED MOMENT

Fig. 2.9 — Strut with eccentric axial load.

$$M = -EI\frac{d^2v}{dx^2}$$

hence equation (2.18) becomes

$$Nv - EI\frac{d^2v}{dx^2} = 0 \tag{2.19}$$

In relation to equation (2.19) note the following:

- The moment Nv tends to further bend the strut
- The moment EId^2v/dx^2 is an internal restoring moment which tends to straighten the strut. Up to a certain load (the buckling load) it will achieve this.
- Taking the curvature as d^2v/dx^2 instead of the correct value of

$$\frac{d^2v/dx^2}{\left[1 + \left[\dfrac{dv}{dx}\right]^2\right]^{1.5}} \tag{2.20}$$

limits the validity to small displacements, i.e. to situations where $(dv/dx)^2$ is small compared with 1.0.

- When equation (2.19) is satisfied, the applied moment and the restoring moment are equal and opposite and the system is in a state of unstable equilibrium. N cannot be increased beyond the buckling value while the small deformation criterion is maintained.

If the small displacement assumption is relaxed and the correct expression for curvature is used then the applied load can increase beyond the buckling value of equation (2.18) provided the elastic limit is not exceeded.

If there are applied loads and/or support reactions causing bending moments the equilibrium equation becomes

$$Nv - EI\frac{d^2v}{dx} + M_x = 0 \tag{2.21}$$

where M_x is the bending moment due to the applied loads and/or support reactions — see Fig. 2.9(c).

A more general form of equation (2.20) is

$$Nv - EI\left[\frac{d^2v}{dx^2} - \left[\frac{d^2v}{dx^2}\right]_0\right] + M_x = 0 \tag{2.22}$$

where $\left[\dfrac{d^2v}{dx^2}\right]_0$ is the initial curvature (i.e. the curvature before N is applied).

Further discussion of non-linear geometry situations is given in section 3.11, 4.8.3 and 8.8.

2.5 SOIL BEHAVIOUR

The behaviour of soil is profoundly affected by the particle size distribution and by the water content. Predictive models are not generally available.

2.5.1 Winkler model
The only possibility for treating the soil using conventional software is to use the Winkler model. Fig. 2.10(a) illustrates this. A pressure q applied over an area A causes uniform deformation over the area but there is no other soil deformation.

Thus over the area of the load the soil acts as a simple spring having the load deformation relationship:

$$q = k_s \delta \tag{2.23}$$

where k_s is the spring stiffness or 'modulus of subgrade reaction'. δ is the deformation of the soil surface. Values of k_s are based on plate bearing tests. The concept of using such a spring provides only a gross approximation since:

- The stiffness of a soil measured by means of a plate bearing test depends on the size of the plate and an extrapolation from test results to a foundataion size is needed.
- Soil behaviour is often dominated by long term movements to which linear elastic conditions do not apply.
- The deformation is assumed to be confined to the loaded area. This does not allow adjacent loaded areas to interact, causing poor results with strip and raft foundations.

(a) WINKLER DEFORMATION   (b) ELASTIC HALF SPACE DEFORMATION

Fig. 2.10 — Soil models.

The main value of the Winkler model is to provide a flexibility for the soil which can be used to study the relative effect of soil movements on structural behaviour rather than as a means of predicting these movements.

A beam on elastic foundation model can be used where the governing diffential equation is

$$EI \frac{d^4 v}{dx^4} + k_s B v = 0 \tag{2.24}$$

where EI is the bending stiffness of the beam, v is the lateral deflection, B is the width of the beam.

Further discussion on the Winkler model is given in sections 3.9.1 and 4.5.3.

Typical values of k_s are given in Table A4.

2.5.2   Elastic half space

A better concept is the elastic half space model. This treats the soil as a linear elastic medium — see equation (2.6) - requiring conventional elastic constants to define its properties. Fig. 2.10(b)shows the deflected profile due to an applied pressure q. This approach has been developed to take account of variations in soil stiffness and non-linearity as discussed in section 3.9.2.

Typical elastic constants for soils are given in Tables A5 and A6.

3

Element behaviour

The 'elements' described in this Chapter are finite parts of the structure whose properties are described in relation to forces and displacements at 'nodes' on the element. The element nodes correspond to nodes on the structure. Degrees of freedom (section 8.2) are defined at the nodes.

A new breed of 'finite' elements was introduced in the 1950s. Prior to that time the basic elements used for structural modelling were 'bar' elements with only axial deformation and 'beam' elements which include axial, bending and torsional behaviour. (In this classification a bar element is a subset of a beam element).

It is useful to define two classes of element:

- Type A elements which do not require mesh refinement. Only uniform bar and uniform beam elements are in this class although these elements are the most widely used in structural modelling.
- Type B elements which do need mesh refinement to improve the accuracy of the solution. The reason why mesh refinement is needed is explained later.

The finite element concept was developed for type B elements but this term is now applied to both type A and type B elements.

There is now a very large number of elements in use. In this chapter the basic assumptions used for some of the simpler and more commonly used elements are discussed. The elements were chosen for description mainly to illustrate points in element formulation rather than to attempt to give a comprehensive view of available elements

3.1 BASIC BEHAVIOUR
3.1.1 Basic assumptions

Element properties are defined in the form of an element stiffness matrix **k** where

$$\mathbf{p} = \mathbf{kd} \qquad (3.1)$$

where **p** is the vector of nodal forces, **d** is the vector of corresponding nodal displacements.

To establish an element stiffness matrix, four main assumptions are needed:

(1) The geometry needs to be defined, e.g. the element can be defined as a straight line, a flat straight-sided quadrilateral, a tetrahedron with parabolic shaped sides, etc.

(2) The degrees of freedom are chosen. This choice depends on the accuracy required from the element and other factors. A definition of the term 'degree of freedom' is given in section 8.2.

(3) The 'constitutive' relationship (Chapter 2) is selected. For example, plane stress or thin plate bending behaviour may be assumed.

(4) Assumptions are made regarding the state of deformation and/or stress within the element.

Most of the elements discussed in this chapter use displacements for assumption 4. These are expressed in terms of either a displacement function where the independent variables do not have a physical interpretation or as shape functions where the independent variables are the nodal displacements. In this section displacement functions are used since they tend to be better for promoting understanding. The shape function approach is normal in programming.

All finite element stiffness matrices can be formulated on the basis of these 4 types of assumption as described in sections 8.5.2 and 8.5.3. For type A elements, other approaches to element stiffness matrix formulation are possible as described in section 8.5.1.

3.1.2   Convergence of type B elements

A main concept in the use of type B elements is that, as the mesh is refined, the result should 'converge' to the 'exact' value (section 1.3). Fig. 3.1 shows typical convergence curves.

Elements based on assumed displacements

When the element chosen involves the use of displacement functions, the displacements within the element will normally be compatible. When the boundaries of adjacent elements fit together in a deformed state, i.e when there is also boundary compatibility, then the element is said to be 'conforming' or 'compatible'. Where boundary compatibility is not satisfied then the element is 'non-conforming'.

A mesh of conforming elements will in general give stiffer predictions than would result from an exact solution. An 'influence coefficient' (i.e. the deflection under a single load) will give a lower bound value and the solution will converge 'monotonically', as illustrated in Fig. 3.1.

A typical non-conforming element (such as the ACM element described in section 3.6.1) assumes element displacements which do not ensure full boundary compatibility. The basic assumption of assuming a displacement function to model the element behaviour restricts the displaced shape of the element and results in the element itself being stiffer than a piece of the constitutive material. A lack of full boundary compatibility causes a relaxation in the stiffness of the element model. Thus compensating assumptions can result in faster convergence for non-conforming elements, as shown in Fig. 3.1.

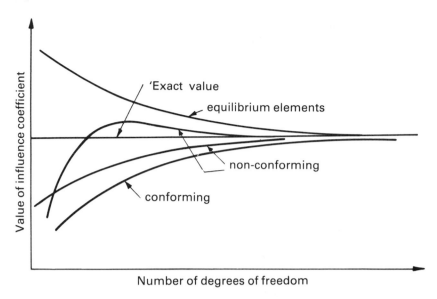

Fig. 3.1 — Convergence curves.

Elements based on assumed stresses

When the element behaviour is based on assumed stresses (which should be in equilibrium within the element) the element itself will be more flexible than a piece of the constitutive material. If equilibrium is also guaranteed across the element boundaries then the mesh of elements will give an upper bound to an influence coefficient and the solution will in general be more flexible than that resulting from an exact solution. Equilibrium elements are not in common use.

Hybrid elements

Hybrid elements (Pian 1966, Walder and Green 1981) use a combination of stress and deformation assumptions and are therefore non-conforming. Typically stresses are assumed within the element and displacements are assumed on the boundaries. The two types of assumption for one element compensate each other and can give very good results. Convergence can be from the stiff or from the flexible side.

Conditions for element convergence

The main requirements for an element to converge to the exact solution is that it should allow rigid body movements and constant strain states. It is argued that if constant strain is admissible then by taking smaller and smaller elements the set of constant strain values can closely approach the sought result.

The Patch Test is a simple means of assessing this. A small number of elements are assembled usually in a square or rectangular shape as shown in Fig. 3.2. To test this system for rigid body movements, such a movement is applied to the boundary nodes but not to the internal nodes. If no significant stresses result in the elements then the rigid body criteria is successfully tested. For the constant strain condition, boundary loading is applied which is consistent with a state of constant stress in all the elements. This can be tested for all stress components individually.

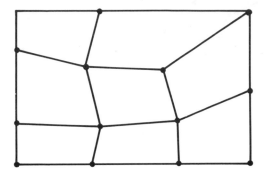

Fig. 3.2 — A typical layout of elements for the patch test.

Benchmark tests

The National Agency for Finite Element Methods and Standards (for address see NAFEMS 1986) publish useful benchmark tests. These are problems for which exact solutions are available and are used to test and compare element performance. When purchasing finite element software, it is good policy to ask for results of such tests.

3.2   JOINT ELEMENTS

A joint element is normally comprised of a set of springs which connect two nodes or two freedoms. The springs are characterized by relationships of the form:

$$p = k\delta$$

where p is a direct force or moment, k is a spring stiffness, δ is the movement in the direction of p.

Up to three translational springs and three rotational springs can be included and various stiffness properties are used.

Fig. 3.3 shows a joint element for in-plane conditions. Both nodes would normally be in the same position but this is not a necessary condition.

Fig. 3.4(a) shows a linear elastic spring relationship. This is used, for example, to model partial foundation fixity conditions. Fig. 3.4(b) shows elasto-plastic behaviour for a rotational spring. Such a characteristic is used to model plastic hinges in frame structures. Another useful model is the contact spring which can for example take compression but not tension. Fig. 3.4(c) shows a more general form of contact spring relationshiop which has a low (or zero) stiffness in one direction and a gap.

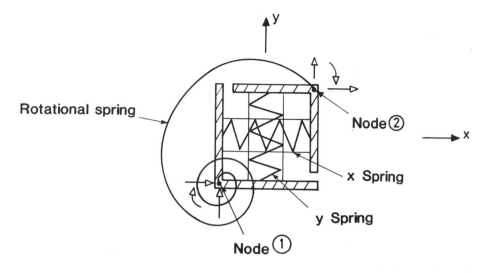

Fig. 3.3 — In-plane joint element.

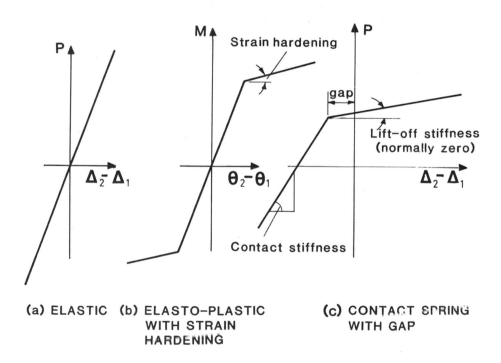

(a) ELASTIC **(b) ELASTO–PLASTIC
WITH STRAIN
HARDENING**

**(c) CONTACT SPRING
WITH GAP**

Fig. 3.4 — Joint element characteristics.

3.3 BAR ELEMENTS

Fig. 3.5 shows a uniform bar element. The assumptions are as follows:

- Geometry — the bar is straight and lies on the local x-axis. It has uniform cross-section throughout its length.
- Degrees of freedom — there is an axial translational freedom at each end of the bar.
- Constitutive relationship:

$$N = EA \ du/dx \tag{3.2}$$

where N is the axial load in the bar, EA is the axial stiffness, u is the deformation in the x-direction.

- Internal displacements

$$u = \alpha_1 + \alpha_2 x \tag{3.3}$$

where α is a coefficient

The number of terms in the displacement function is normally equal to the number of degrees of freedom for the element. The development of the stiffness matrix for this element based on the above assumptions is given in section 8.5.2.

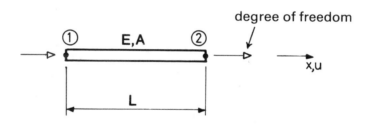

Fig. 3.5 — Uniform bar element.

Differentiating equation (3.3) gives $du/dx = \alpha_2$, i.e. constant strain along the length of the bar. Since N is constant along the length of the element, the constitutive relationship — equation (3.2) — also infers constant strain and therefore the displacement assumption gives a correct solution for the constitutive equation. Therefore this is a type A element which does not require mesh refinement.

The uniform stress assumption is good for bars in tension where the ends are pinned and/or the bar has low lateral stiffness (e.g. in the case of a tie rod). In compression, the assumption can be acceptable for 'short' columns with negligible eccentricity of the applied load. With 'long' columns the effect of eccentricity of the applied load needs to be taken into account as discussed in section 3.11.

The bar element can readily be transformed into two dimensions (having two translational freedoms per node) or to three dimensions (having three translational freedoms per node).

Other bar elements are available which have variable cross-section and a mid-side node (e.g. Lusas 1987). These are type B elements.

3.4 BEAM ELEMENTS

The term 'beam' tends to denote bending but in the context of structural elements it normally refers to a general line element which has bending, shear, axial and torsional deformation. This description is used here.

In this section the components that make up beam elements are briefly discussed. Some derivations and element matrices are given in section 8.5.1.

3.4.1 Bending

Engineer's Theory of Bending is possibly the most commonly used structural model. It gives remarkably good results in a wide range of situations but its behaviour on the margins of accuracy is probably not well understood.

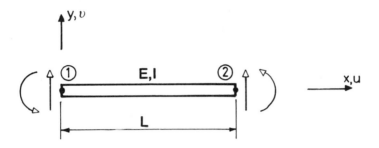

Fig. 3.6 — Uniform bending element.

Fig. 3.6 shows a uniform in-plane bending element. The assumptions for this element are as follows:

- Geometry — the element has uniform cross-section throughout its length. It lies on the (local) x axis of plane $x - y$.
- Degrees of freedom – there is a translational and a rotational freedom at each end of the element.
- Constitutive relationship — see equation (2.7).
- Internal displacements:

$$v = \alpha_1 + \alpha_2 x + \alpha_3 x^2 + \alpha_4 x^3 \tag{3.4}$$

From the above assumptions the 4×4 element stiffness matrix can be formulated.

This is another type A element since differentiating equation (3.4) twice gives

$$d^2v/dx^2 - 2\alpha_3 + 6\alpha_4 x \tag{3.5}$$

With only end actions on the element as shown in Fig. 3.6 the bending moment in the element is linear and therefore equation (2.7) infers that the curvature should be linear. Equation (3.5) shows that the curvature is linear and therefore equation (3.4) is a correct solution for equation (2.7).

3.4.2 Bending with shear deformation

A method of incorporating shear deformation into a bending element stiffness matrix is described in section 8.5.1. When shear deformation is included, the end rotation is interpreted as the rotation of the cross-section which is normal to the longitudinal axis of the beam before load is applied rather than the slope of the longitudinal axis — see Fig. 8.7(b).

Bending and shear deformations modes

The distinction between bending mode and shear mode deformation in beams is important both for understanding of beam behaviour and for the use of simplified models of frames and walls discussed in Chapter 6.

Fig. 3.7(a) shows a rigid block and spring model of a tip loaded cantilever demonstrating bending mode behaviour. Note that the rotation of the blocks has a magnifying effect on the tip deflection.

Fig. 3.7(b) gives a similar model for shear mode deformation. In this case the tip deflection is the sum of the relative movements of the blocks and with constant shear, as in this case, the deflected shape is a straight line.

Equation (2.10) is the constitutive relationship for shear deformation, i.e.

$$S = \overline{A}G \ dv/dx$$

If equation (2.7) is differentiated with respect to x we get the bending equivalent of equation (2.10), i.e.

$$dM/dx = S = EI \ d^3v/dx^3 \tag{3.6}$$

Comparing equations (2.10) and (3.6) shows that for a given loading the bending mode shape is two orders higher than the shear mode shape. Further information about shear deformation of beams is given in section 8.5.1 and Appendix 2.

Case study 3.1   Beam bending

In this example the behaviour of beams at normal and low span to depth ratios, the relative effect of shear deformation and the effect of different support assumptions are investigated.

Model type S (steel) is a simply supported steel I beam taking a uniformly distributed load as shown in Fig. 3.8(a). Model type C (concrete) has the same supports and loading but is of concrete with a rectangular cross-section as shown in Fig. 3.9(a). Span to depth ratios of 5:1 and 2:1 are considered and support conditions with and without a roller are included.

In each case a plane stress element model and a beam element model are used.

Figs 3.8(b) and 3.9(b) show the plane stress meshes. A symmetrical half was used with boundary conditions as shown. The models shown in Figs 3.8 and 3.9 all show the roller support case. In each case a separate run with a pinned support was produced. Quadrilateral hybrid elements with 12 degrees of freedom per element (Walder and Green 1981) were used to model the web. Beam elements with bending and axial deformation were used for the flanges (flange bending would be negligible). The load was applied at the top flange level. The material properties are given in Figs 3.8(a) and 3.9(a).

Figs 3.8(c) and 3.9(c) show the 2:1 aspect ratio model and Figs 3.8(d) and 3.9(d) are the beam element models. The rigid arms which simulate the finite depth of the beam in the beam element model only affect the results when the support is pinned.

Table 3.1 gives selected results. Note that neither the beam element model nor

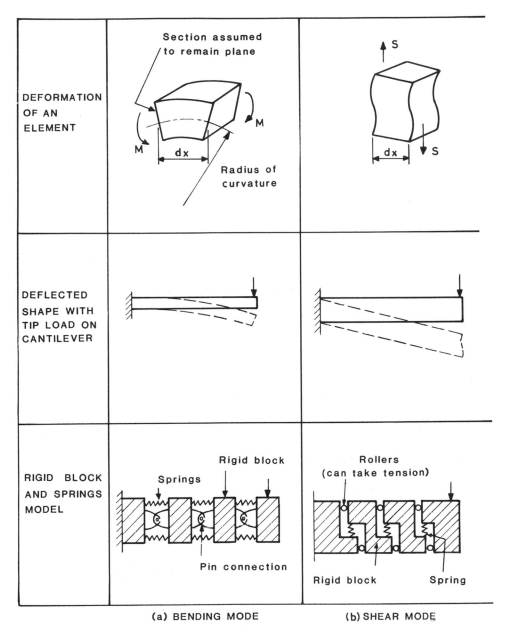

(a) BENDING MODE   (b) SHEAR MODE

Fig. 3.7   Bending and shear modes.

the plane stress model necessarily compare closely with the exact solution (see section 1.3). The beam element is probably better for the 5:1 span ratio and the plane stress elements are likely to be better for the 2:1 span ratios.

Note the following:

- With 5:1 span ratio the plane stress models and the beam models give results that are fairly close to each other.

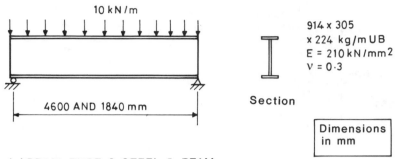

914 x 305
x 224 kg/m UB
E = 210 kN/mm²
v = 0·3

Section

Dimensions
in mm

(a) BEAM TYPE S-STEEL I BEAM

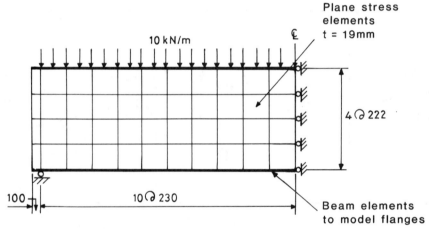

(b) PLANE STRESS MODEL FOR 4600 mm SPAN

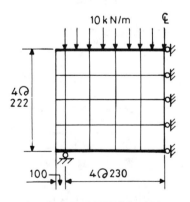

(c) PLANE STRESS MODEL
FOR 1840 mm SPAN

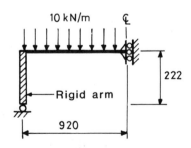

(d) BEAM ELEMENT MODEL

Fig. 3.8 — Beam example — steel beam.

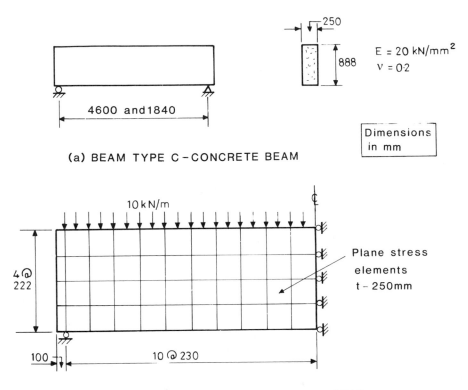

E = 20 kN/mm²
ν = 0·2

Dimensions in mm

(a) BEAM TYPE C – CONCRETE BEAM

(b) PLANE STRESS MODEL FOR 4600mm SPAN

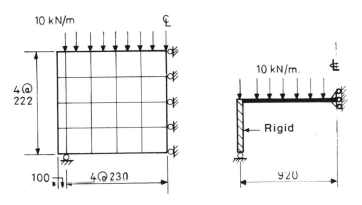

(c) PLANE STRESS MODEL   (d) BEAM ELEMENT MODEL
FOR 1840mm SPAN

Fig. 3.9 — Beam example — concrete beam.

- The beam models with a roller have a horizontal axis of symmetry whereas the corresponding plane stress models do not since they take account of the fact that the load is applied at the top of the beam.
- The effect of shear deformation is much more pronounced with the I section models. This is because the flanges significantly reduce the bending deformation but only marginally reduce the shear deformation.
- Bending theory gives only approximate results for the 2:1 aspect ratio cases (assuming that the plane stress results are more accurate in these situations)
- Whether or not there is a roller support makes a significant difference to the results in most cases. It seems to have less effect on the deeper beams than one would expect. This is because these models are controlled by shear deformation which is not affected by the pin restraint.

Much insight can be gained by critical study of the results in Table 3.1.

Table 3.1 — Case study — beam bending

Model No.	Span/ depth	Support	Model type	Δ_{max} mm	σ N/mm$^2 \times 10^3$		Δ Shear/ Δ Bending
					Top	Bottom	
S1		Roller	PS	1.003	− 28.2	29.9	
S2	5:1		Beam	0.964	− 31.0	31.0	0.23
S3		Pin	PS	0.671	− 25.1	9.93	
S4			Beam	0.611	− 27.2	10.43	
S5		Roller	PS	0.0949	− 3.74	5.53	
S6	2:1		Beam	0.0550	− 5.01	5.01	1.91
S7		Pin	PS	0.0796	− 3.11	1.57	
S8			Beam	0.0460	− 4.34	1.68	
C1		Roller	PS	2.20	− 7.90	7.90	
C2	5:1		Beam	2.17	− 8.05	8.05	0.087
C3		Pin	PS	1.11	− 5.64	3.27	
C4			Beam	0.972	− 5.37	2.68	
C5		Roller	PS	0.109	− 1.31	1.34	
C6	2:1		Beam	0.0787	− 1.29	1.29	0.539
C7		Pin	PS	0.0758	− 0.594	0.927	
C8			Beam	0.0480	− 0.859	0.429	

PS — Plane stress. Values quoted are at centre line.

3.4.3 Torsion

A St Venant torsion component is normally included in a beam element as described in section 8.5.1. This has end torques and end rotation as shown in Fig. 3.10(a). A combined torsion element (Reilly 1972) is also described in section 8.5.1 — see Fig. 3.10(b). In addition to end torques the combined element has end bimoments — see section 2.2.9. The deformation corresponding to a bimoment is dφ/dx, i.e. the first derivative of twisting rotation. Such deformation is not easy to represent pictorially. Elements which also include distortion of the cross-section are also available (Lusas 1987).

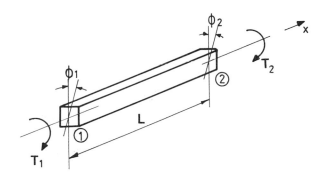

(a) ST VENANT TORSION ELEMENT

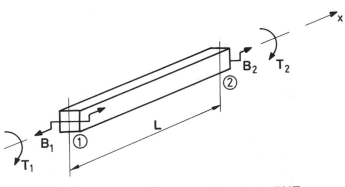

(b) COMBINED TORSION ELEMENT

Fig. 3.10 — Torsion elements.

A simple example of warping torsion is an I beam fully fixed at one end with an applied torque T at the other end — Fig. 2.6(b). The torque can be treated as two point loads T/D acting on the flanges, as shown in Fig. 3.11. The flanges can now be treated as cantilevers with a top point load and stresses and deformations calculated using bending theory.

3.4.4   Beam element types
Plane frame element
By combining a bar element (Fig. 3.5) and a bending 'element (Fig. 3.6) one obtains a plane frame beam element — Fig. 3.12(a). This is one of the most commonly used

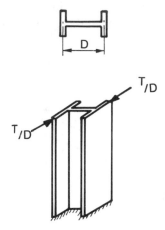

WARPING TORSION OF AN I SECTION

Fig. 3.11 — Warping torsion of an *I* beam.

elements. The axial and bending components are normally assumed to be independent but non-linear geometry effects can be considered as discussed in section 3.11.

Grillage element
By combining a torsion element and a bending element one obtains a grillage element — Fig. 3.12(b).

Space frame element
By combining a bar element, bending elements for bending about both the *x*- and *y*-axes and a torsion element one obtains a space frame element — Fig. 3.12(c). This has six degrees of freedom per node. If the warping torsion terms are also included there are seven degrees of freedom per node.

3.4.5   Member loading
Type A elements are only directly valid for end loading but this is overcome by using the concept of joint loads. This requires the superposition of two load cases as illustrated for a two span continuous beam in Fig. 3.13. Load case 1 has restraining member end moments which cause the rotations at the degrees of freedom to be zero — Fig. 3.13(b). The opposites of these restraining moments are then applied at the structural freedoms as shown in Fig. 3.13(c). The final result is the sum of the results of the two load cases.

3.5   MEMBRANE ELEMENTS

Membrane elements have properties defined in a plane. They are all type **B** elements.

In this section the basic assumptions for some typical membrane elements are described to help develop an insight into element behaviour.

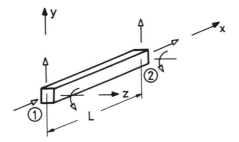

(a) PLANE FRAME ELEMENT

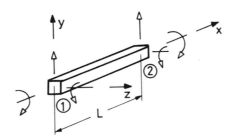

(b) GRILLAGE ELEMENT

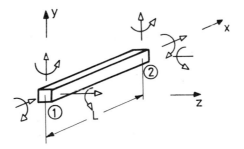

(c) SPACE FRAME ELEMENT

Fig. 3.12 — Beam element types.

3.5.1   The triangular constant strain element (CST)

The assumptions are:

- Geometry – as its name implies, it is a flat straight-sided triangle (Fig. 3.14).
- Degrees of freedom — there are two translational freedoms at each node giving a total of six freedoms per element (Fig. 3.14).

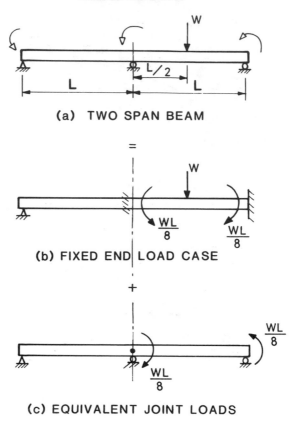

(a) TWO SPAN BEAM

(b) FIXED END LOAD CASE

(c) EQUIVALENT JOINT LOADS

Fig. 3.13 — Treatment of member loads for beam elements.

● Constitutive relationships — either the plane stress or the plane strain constitutive relationship as given in section 2.2.1 or 2.2.2 can be used.
● Displacement functions are:

$$u = \alpha_1 + \alpha_2 x + \alpha_3 y$$

$$v = \alpha_4 + \alpha_5 x + \alpha_6 y \qquad\qquad (3.7)$$

where u and v are the displacements in the x- and y-directions respectively and α_i is a constant.

   Note that the direct strains are:

$$\frac{\partial u}{\partial x} = \alpha_2, \frac{\partial v}{\partial y} = \alpha_6$$

and the shear strain is:

$$\frac{\partial u}{\partial y} + \frac{\partial v}{\partial x} = \alpha_3 + \alpha_5$$

Thus all strain (and hence stress) components over the area of an element are

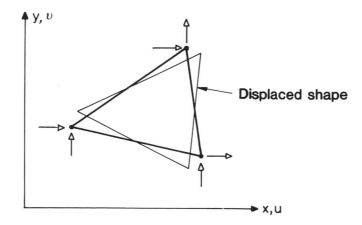

Fig. 3.14 — Constant strain triangle.

constant. This then is the main assumption made as to the behaviour of the element. Displacements are linear and strains are constant. For example, if a beam in bending is modelled using such elements the resulting stress distribution at a section will be as shown in Fig. 3.15. Instead of being close to linear the stress will be stepped. However, if sufficient elements are used, the stepped distribution will approach the correct one. This is the basis of the use of type B elements. Although the assumptions made regarding the distribution of deformation and stress within the element may be crude, use of sufficient elements will give results to an acceptable accuracy.

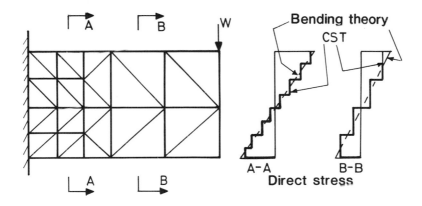

Fig. 3.15 — Stress prediction of a beam using constant strain triangle.

This is a conforming element (section 3.1.1) and is one of the simplest possible two-dimensional elements. Most other membrane elements have more degrees of freedom and are more accurate.

3.5.2 Four-noded linear strain quadrilateral element

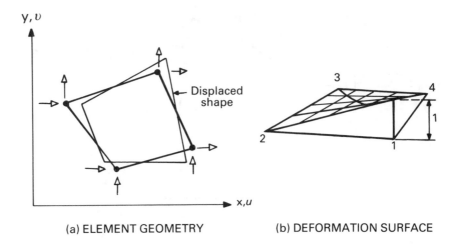

(a) ELEMENT GEOMETRY   (b) DEFORMATION SURFACE

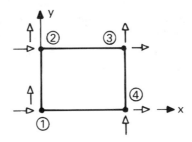

(c) RECTANGULAR FORM

Fig. 3.16 — Quadrilateral plane stress element with four nodes.

The assumptions are:

- Quadrilateral shape in the x-y plane — Fig. 3.16(a).
- Degrees of freedom — two translational degrees of freedom at each corner, giving a total of eight freedoms per element.
- Constitutive relationships — either plane stress or plane stain as described in section 2.2.1 or 2.2.2.
- The displacement functions are:

$$u = \alpha_1 + \alpha_2 x + \alpha_3 y + \alpha_4 xy$$
$$v = \alpha_5 + \alpha_6 x + \alpha_7 y + \alpha_8 xy \tag{3.8}$$

Using these assumptions the shape of the displacements u and v over the area of the rectangular form of the element is a hyperbolic parabola (Fig. 3.16(b)). Along a line parallel to the x or y-axes the displacement (u or v) is linear, and along a diagonal line across the element the displacement is hyperbolic.

Boundary compatibility
Consider the rectangular form of the element shown in Fig. 3.16(c). Along the boundary $x = 0$ the u displacement is $\alpha_1 + \alpha_3 y$, i.e. it is linear. This displacement is defined by the u displacements at nodes 1 and 2. Adjoining elements on line 1–2 will have the same u displacement on this line. Similar arguments for the v displacements and for the other element sides demonstrate that this is a conforming element.

This is an 'isoparametric' element, so called because the same functions can be used to describe the geometry as define the displaced shape — section 8.5.4. The nodal coordinates are the independent variables for the element shape and the corresponding displacements are the independent variables for the deformed shape. If the functions uniquely define the element edges they also uniquely define the edge displacements. Therefore isoparametric elements are conforming. The derivation of the element stiffness matrix for this element using shape functions is discussed in section 8.5.4.

3.5.3 Quadrilateral element with mid-side nodes
The following assumptions are used for this element:
- Geometry — the element has four sides, each side being capable of having a parabolic shape (Fig. 3.17(a)).
- Degrees of freedom — there are four corner nodes and four mid-side nodes each with two translational freedoms giving a total of sixteen freedoms.
- Constitutive relationship — plane stress or plane strain (section 2.2.1 or 2.2.2).
- The displacement functions used are:

$$u = \alpha_1 + \alpha_2 x + \alpha_3 y + \alpha_4 xy + \alpha_5 x^2 y + \alpha_6 xy^2 + \alpha_7 x^3 + \alpha_8 y^3$$
$$v = \alpha_9 + \alpha_{10} x + \alpha_{11} y + \alpha_{12} xy + \alpha_{13} x^2 y + \alpha_{14} xy^2 + \alpha_{15} x^3 + \alpha_{16} y^3 \qquad (3.9)$$

Fig. 3.17(b) shows the parabolic displaced shape of the boundaries. This is another isoparametric and therefore conforming element. It is part of a 'family' of elements (Zienkiewicz 1977) of which the four-noded element described in section 3.5.2 is the most elementary.

This eight-noded element is considered to be a 'good performer' and is widely used. I have a probably unreasonable antipathy to elements with side nodes. They are more cumbersome to use than elements with only corner nodes unless mesh generating facilities can be used but significantly fewer elements are needed.

3.6 PLATE BENDING ELEMENTS
Elements for thin plate bending often have fairly complex formulations. One plate bending — the ACM rectangular element — is described here. It was the first plate

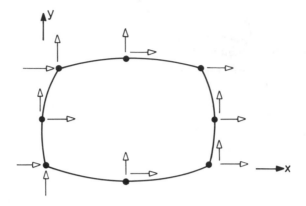

(a) Curved (Parabolic) sides

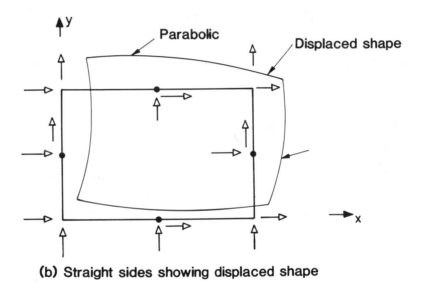

(b) Straight sides showing displaced shape

Fig. 3.17 — Qudrilateral plane stress element with eight nodes.

bending element to be established, being developed independently by Adini and Clough (1961) and Melosh (1963), hence the ACM name. Better plate bending elements are now available but a study of the assumptions for the ACM element is helpful in understanding some important aspects of element formulation.

3.6.1 The ACM rectangular plate bending finite element
The assumptions are:

- Geometry — it is a rectangular flat element.

- Degrees of freedom — at each corner three degrees of freedom are chosen corresponding to a translation at right angles to the plane of the plate and rotations about the x- and the y-axes (similar to the freedoms for a grillage element) — Fig.3.18(a).
- Constitutive relationship — the thin plate bending constitutive relationship — equation (2.8).
- The displacement function used is:

$$w = \alpha_1 + \alpha_2 x + \alpha_3 y + \alpha_4 xy + \alpha_5 y^2 + \alpha_6 x^2 + \alpha_7 x^2 y + \alpha_8 xy^2$$
$$a_9 x^3 + \alpha_{10} y^3 + \alpha_{11} x^3 y + \alpha_{12} xy^3 \tag{3.10}$$

A twelve-term function is needed so it is not possible to use a complete fourth order function which has fifteen-terms (terms in x^4, y^4 and $x^2 y^2$ are missing in the above function). The question arises — why choose one set of terms rather than another?

In this case, two terms from five of the fourth order set need to be discarded, i.e. two from $x^3 y$, xy^3, x^4, y^4, and $x^2 y^2$ must be chosen. The resulting function should be symmetrical and therefore the $x^2 y^2$ term must be discarded and the choice falls on $x^3 y$ and xy^3 or x^4 and y^4. As we will see, using the x^4 and y^4 terms would cause inter-element boundary compatibility to be non-existent. Therefore the $x^3 y$ and xy^3 terms are used.

Equation (3.10) is in fact the Lagrange interpolation function in two dimensions. In Melosh's original paper (Melosh 1963) Lagrange shape functions are used in the element formulation. Early workers were confused by typographical errors in this paper which made his element (derived using the shape function approach described in section 8.5.4) apparently different from the Adini–Clough version.

Boundary compatibility

To examine the inter-element compatibility for this element consider Fig. 3.18(b) which shows two ACM elements which form part of a mesh. Section AA — Fig.3.18(c) — is on a line $x = $ constant. If $x = $ constant is substituted into equation (3.10) the displacement w on line cd will have the form

$$w = c_1 + c_2 y + c_3 y^2 + c_4 y^3 \tag{3.11}$$

where c_i is a constant.

On line cd there are four nodal deformations relevant to equation (3.11), i.e. w_c, θ_{xc}, w_d, θ_{xd} as shown in Fig. 3.18(c). These four deformations are sufficient to uniquely define the cubic function of equation (3.11). Therefore adjacent element boundaries have the same transverse deflection along lines parallel to the y-axis. Since equation (3.10) is symmetrical, the same arguments holds for boundaries on lines parallel to the x-axis. Note that the argument would not hold if the x^4 and y^4 terms had been used in the displacement function.

Now consider the cross slope on a $x = $ constant boundary. This has the form:

$$\theta_y = \partial_w / \partial_x = c_1 + c_2 y + c_3 y^2 \tag{3.12}$$

In this case there are only two nodal values to define the θ_y function (i.e. the rotations θ_{yc} and θ_{yb}) where three are required. Therefore the element involves a discontinuity of cross slope at interelement boundaries — Fig. 3.18(d) — and is non-conforming.

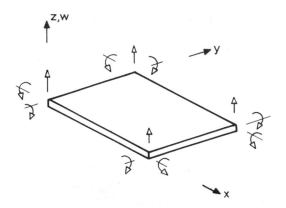

(a) 12 DEGREE OF FREEDOM PLATE BENDING ELEMENT

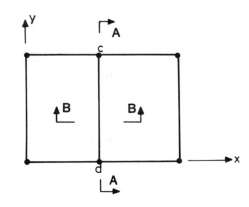

(b) 2 CONNECTED ELEMENTS

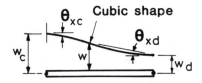

(c) SECTION A–A SHOWING DISPLACED SHAPE OF BOUNDARY

(d) SECTION B—B SHOWING CROSS SLOPE DISCONTINUITY

Fig. 3.18 — ACM plate bending element.

3.6.2   Other plate bending elements

Several types of triangular plate bending elements have been formulated. Some of them do not perform well and in general it is better to avoid such elements by using quadrilateral shapes. Advice on how this can be achieved is given in section 4.1.2.

Much effort has gone into developing plate bending elements. Some elements have had their terms adjusted to get better matching with accurate solutions and there is no clear guide to the best element for use.

For thicker plates one can use the Mindlin constitutive relationship (Mindlin 1951) which is the biaxial bending equivalent of shear deformation of beams (section 3.4.2).

3.7   SHELL ELEMENTS

By combining a membrane element and a plate bending element one obtains a flat 'shell' element — Fig. 3.19(a). This is a useful form of element for slab analysis where membrane effects are not negligible. It is not suited for curved shell analysis and, in fact, cannot be fitted to a doubly curved surface. A triangular facet shell element (Fig. 3.19(b)) can be used more readily for curved shells.

The elements shown in Fig. 3.19 do not show in-plane rotations at the nodes. Most plane stress elements do not have such freedoms (see, however, Walder & Green 1981) so transformation to a system of global coordinates with six degrees of freedom requires special provisions, especially when all the elements at a point are in the same plane (Zienkiewicz 1977).

For curved surfaces, curved elements are more appropriate. Many such elements are available. A commonly used curved shell element is the semi-loof element (Irons 1976). The formulation of this element is complex and indeed Irons & Ahmed (1980) state: 'We are convinced that few workers will be motivated to enquire in any depth [about the semi-loof shell element]. Indeed, we ourselves now find the original reference logical but extremely taxing to read.'

I find this statement to be most revealing. Many, if not most, finite elements are understood in terms of how they perform and not in terms of how they are derived.

3.8   SOLID ELEMENTS

Six noded 'tetrahedron' elements or eight-noded 'brick' elements — Fig. 3.20 — are used to model 'thick' structures. A minimum of three translational degrees of freedom per node are needed (i.e. a total of twelve for a tetrahedron and twenty four for a brick) but 'higher order' elements with more than this number of freedoms are also available. Extra freedoms are either translations at extra nodes or correspond to derivatives of nodal displacements.

A special case of solid element is the axisymmetric element which has a toroidal shape, the freedoms being defined only in relation to one cross section.

Solid elements are seldom used in structural modelling. Normally the behaviour of systems that need to be modelled in three dimensions is complex and non-linear material properties are needed. Solutions for non-linear 3D problems can be very expensive.

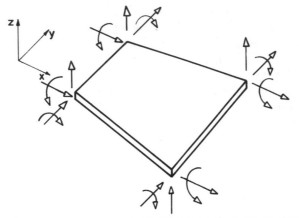

(a) QUADRILATERAL FLAT SHELL ELEMENT

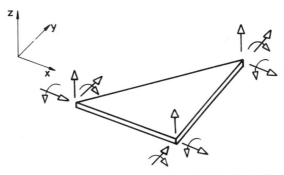

(b) TRIANGULAR FACET ELEMENT

Fig. 3.19 — Flat shell elements.

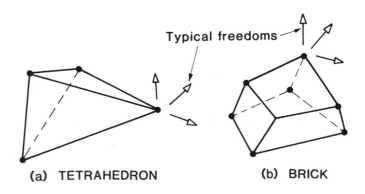

Typical freedoms

(a) TETRAHEDRON   **(b) BRICK**

Fig. 3.20 — 3D elements.

3.9 ELEMENTS TO REPRESENT SOIL BEHAVIOUR
3.9.1 Winkler elements

The basis of the Winkler concept for soil deformation is discussed in section 2.4. There are two ways in which a Winkler support can be introduced.

Winkler springs

Although a Winkler support can be treated as a continuous function its simplicity (and main drawback) lies in the fact that there is no effective coupling of deformations via the soil. Therefore a set of support springs can be used to model a Winkler support. Fig. 3.21(a) shows a beam on such a support and Fig. 3.21(b) shows the spring support equivalent. The springs are given a stiffness k_{sp}, where

$$k_{sp} = k_s b_s \tag{3.13}$$

where b is the width of the beam, and s is the spacing of the springs.

As s decreases one approaches the result given if the support is treated as continuous. In view of the high degree of approximation in the Winkler model, a close spacing of the springs is not necessary.

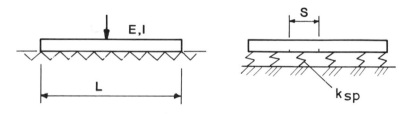

(a) Beam on Winkler support   (b) Spring support model

Fig. 3.21 — Winkler models.

Beams and plates on Winkler supports

For a beam, equation (2.24) can be solved for different load cases. Hetenyi (1946) is the standard work on this subject.

A stiffness matrix for a beam element on a Winkler support can be established by integrating equation (2.24). A Winkler support term can also be added to a plate bending element stiffness relationship (Walder & Green 1981).

3.9.2 Elastic half space elements

A better soil model is provided by using elastic half space elements. Fig. 3.22 shows a plan of a soil surface on which is defined a grid of nodes. At a typical internal node — say node 6 — a load P6 is considered to act over the area $a \times b$. The vertical deformation at node 6 due to this load and at all other nodes is calculated to give a vector of flexibility coefficients. These deformations are calculated by integrating the strains resulting from the Boussinesq stress distribution (Cheung & Zienkiewicz 1965). This is repeated for all internal nodes.

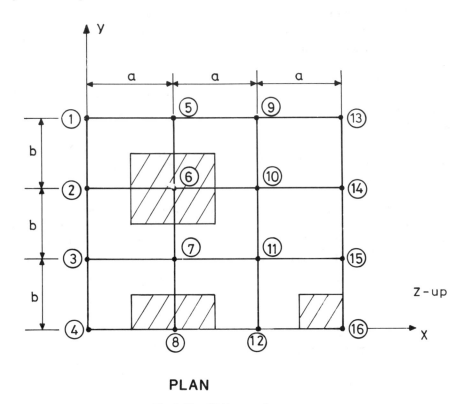

PLAN

Fig. 3.22 — Half space element.

A similar calculation is carried out for the side nodes using a corresponding tributary load. For example, for node 8 the tributary load is $axb/2$.

For the corner nodes the tributary area is $axb/4$.

The vectors of flexibility coefficients thus calculated are put together to form a flexibility matrix which is then inverted to produce a stiffness matrix. This can be treated as an element stiffness matrix although it is somewhat different from most conventional elements in having a grid of nodes rather than mainly side and corner nodes. At each node of the element there is a single vertical degree of freedom. This stiffness relationship can be modified to take account of layering in the soil. One way that this is done is to assume that the Boussinesq stress distribution is not affected by the change in stiffness. These stresses are converted to strains based on the layered properties and integrated to calculate the surface displacements (Wood 1977).

3.10 ELEMENTS WITH NON-LINEAR MATERIAL BEHAVIOUR

To introduce non-linear material behaviour, a suitable alteration to the constitutive relationship is made at step 3 of the basic assumptions of section 3.1.1. If this is done one needs to consider the following factors:

- The processing of the equations is iterative and requires a lot more computing time than for linear problems.
- Convergence to a solution can present problems.
- The user needs more experience to deal successfully with such problems than for linear problems.

It is as yet not common to use non-linear analysis in structural engineering practice but it is likely that this type of modelling will become more widely used as computing facilities become cheaper and solution routines become more robust.

Subject to the factors listed above, the scope for obtaining solutions to non-linear problems is relatively unlimited. Technology for obtaining solutions has outstripped our ability to define the behaviour of engineering materials. This divergence is likely to continue as computing power improves.

Use of non-linear models is discussed in section 4.8.

3.11   ELEMENTS WITH NON-LINEAR GEOMETRY BEHAVIOUR

The effect of magnification of moment in a beam element caused by eccentricity of the axial load resulting from the bent shape of the element can be incorporated into a beam element stiffness matrix. This is done in one of two basic ways.

Equation (2.19) can be solved for given boundary conditions and a stiffness matrix established, the terms of which involve circular functions normally called 'stability functions'.

Alternatively, the stiffness matrix can be established on the basis of a cubic displacement function using the techniques described in section 8.8.1. This is equivalent to taking the first two terms of a series expansion of the stability functions previously mentioned — see Allen & Bulson (1980). The resulting stiffness matrix can then be expresssed as the sum of the conventional linear elastic matrix plus a 'geometric stiffness matrix' which takes account of the non-linear geometry effect.

The geometric stiffness matrix approach is normally used for programming because it is easier to implement than the stability function method. However, the degree of approximation in taking only two terms of the series expansion becomes unacceptable as the axial load approaches the Euler critical load $N_E = \pi^2 EI/L^2$. Allen & Bulson (1980) advise that if N/N_E exceeds 0.4 then the member should be subdivided by adding nodes within the length of the element. For struts, two extra nodes would normally be sufficient.

4

System models

This chapter describes techniques for setting up models of structural systems using conventional analysis packages. Simplified models for which hand analysis is feasible are discussed in section 6.

The first step in modern analytical modelling of a structure is to define a set of elements with support conditions and loading so as to represent the prototype to an adequate degree. The advice given in this section on how to do this is not intended to be prescriptive. One has to decide in a given situation what will be adequate and there is much scope for inventiveness.

Concepts used in structural design have been restricted by our analytical ability and we still cling to traditional structural forms devised before the computer era. Our greatly improved analytical facilities should prompt the development of more imaginative structures.

However, a change of analytical approach can have unexpected results and some examples of this are discussed in this chapter.

As computing power becomes more readily available and cheaper it is likely that the use of three-dimensional models will become more common. As yet, however, we still tend to break down our structures into planar components.

Appendix 1 gives useful data for material and element properties.

4.1 GENERAL

4.1.1 Choice of element type

In accordance with Principle 1 in section 1.2, structural systems should be modelled as beam (or bar) elements whenever this is practicable for the following reasons:

- If the beam elements have uniform section then type A elements (section 2.1) which do not need mesh refinement are used. This significantly reduces the number of degrees of freedom required in comparison with using type B elements.

● The output from beam elements is in the form of forces and moments which are suitable for use directly for member sizing.

4.1.2 Mesh refinement for type B elements

Type B elements are those for which mesh refinement is required. When choosing a mesh of such elements the following criteria can be considered:

● Use a finer mesh where the stress gradients are high and a coarse mesh in areas of more uniform stress.

● Choose a mesh which gives a result which is just on the 'flat' part of the convergence curve — see Fig. 4.1.

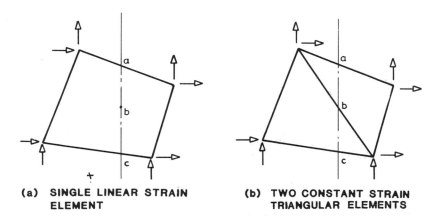

(a) **SINGLE LINEAR STRAIN ELEMENT**   (b) **TWO CONSTANT STRAIN TRIANGULAR ELEMENTS**

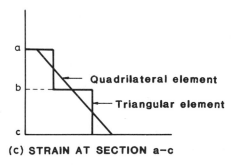

(c) STRAIN AT SECTION a–c

Fig. 4.1 — Strain prediction with a single quadrilateral element or two triangular plane stress elements.

● Keep the number of nodes and elements to a minimum.

● Try to reduce the amount of data input by taking advantage of mesh generating facilities.

- Keep the elements as 'square' as possible, i.e. keep length to breadth ratio as close to unity as possible and angles close to 90° in quadrilaterals and angles close to 60° in triangles. This requirement can be relaxed with some elements — especially higher order elements and where stresses are low and unimportant.
- With linear analysis use quadrilateral elements wherever possible in preference to triangles. Remove diagonal lines to create quadrilaterals from two triangles. For example, at the top right hand corner of Fig. 4.2(d), one could remove two lines to create two quadrilaterals from four triangles.
- At a re-entrant corner the theoretical elastic stresses are infinite and therefore refinement of the mesh in such areas will show very high stresses. In reality there is likely to be localized cracking or plastic flow such that the peak predicted stress may not occur.

The above criteria are not all easy to satisfy and are often competing. One may not know where the high stress gradients are located. A convergence curve for a particular problem is unlikely to be available and using mesh generators can result in more element than are necessary.

Using quadrilateral elements in preference to triangular ones seems paradoxical. Does using more elements not constitute a mesh refinement? In terms of accuracy the answer to this question is 'no'. Triangular elements tend not to perform as well as quadrilateral ones. By combining two triangles to create a quadrilateral, higher order functions can be used to describe the behaviour. This is illustrated in Fig. 4.1. The linear strain at a section is a better representation of the real behaviour than two constant strain values without increasing the number of degree of freedoms.

A logical question then is — 'Would it not be better to use even higher order elements (i.e. elements based on higher order functions? What about a five-sided element?' Such an element would be impractical but higher order quadrilaterals using extra nodes (e.g. mid-side nodes) or extra degrees of freedom at nodes (corresponding to derivatives of nodal deformation) are common.

However, higher order elements require more computational effort to set up and their behaviour is less easy to understand. With non-linear analysis there are significant advantages in keeping the element simple. For example, requiring non-linear material properties to be varied over the area of a higher order element causes difficulties in defining the element behaviour. In situations where there are signifi-cant variations in properties and with time stepping it may be best to use constant strain triangles.

Case study 4.1
Fig. 4.2 shows a cantilever bracket for which the stress in the area of the applied load is to be investigated using plane stress elements. (a) shows a rough mesh with no refinement; (b) is an economical refinement in terms of numbers of elements and degrees of freedom; (c) illustrates a common error made in mesh refinement. At points 'x' there will be a node which relates to the smaller elements but not to the larger ones unless they have mid side nodes or a constraint equation is used to make sure that the boundary of the larger element is compatible with the node. Unless one of these conditions pertains then the refinement technique of (c) should not be used. (d) is a spider web approach which gives good proportioning of elements but may be

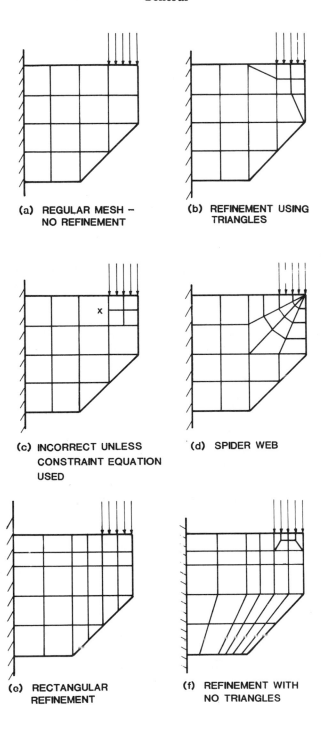

Fig. 4.2 — Mesh layouts for cantilever bracket.

difficult to generate. (e) uses mainly rectangular elements which is helpful for generating the mesh but can result in elements that are long and narrow. In (f) there are no triangular elements but the element shapes are not all satisfactory.

The mesh below the load in Fig. 4.2(f) shows an alternative technique for local refinement.

4.2 FRAMES

The two main components of deformation in the members of frames are axial and bending deformation. In the past, because of our limited ability to create solutions, we tended to separate these components by developing trusses with bending deformation neglected and rigid jointed unbraced frames with axial deformation neglected. The truss approach is satisfactory for triangulated systems with relatively slender members (e.g. roof trusses) and to neglect axial deformation in rigid jointed unbraced frames is normally acceptable. Nowadays there is no need to neglect either effect. In fact, in a computer model it is normally easier to include them than to exclude them.

However, in a truss system, inclusion of bending stiffness may result in significantly larger member sizes and the cost of the connections to provide the moment connections can be important. Principle 3 in section 1.2 is relevant here.

Section 3.4 discusses beam element behaviour and Tables A1, A2 and A3 give section property data. Frame model applications are discussed in other parts of this chapter. In this section some general considerations regarding modelling are discussed.

4.2.1   Finite size of members

The finite size of beam elements can be modelled using rigid offsets. These are created using constraints as described in section 8.1.

Fig. 4.3(a) shows a steel beam supported on columns with moment connections. Fig. 4.3(b) shows a model which takes account of the finite width of the column — denoted as Model A. Fig. 4.3(c) shows the more conventional model with the beam being flexible up to the column centrelines — Model B. Factors to be considered are:

- Due to the rigid offset, the beam of Model A will be stiffer than that of Model B.
- The flexibility of the joints between the beam and the columns will reduce the effective stiffness of the beam and neglecting the finite width of the column would allow for this to some degree.
- The moment at the face of the column is that needed for design of the beam. This is likely to be given as output from Model A which could be useful.

Fig. 4.3(d) shows the beam with rigid ends under reverse bending; a common situation under lateral load on frames. The relationship between end moment — M — and rotation — θ — for his beam is

$$M = \frac{6EI}{L(1-\beta)^3}\theta \qquad\qquad (4.1)$$

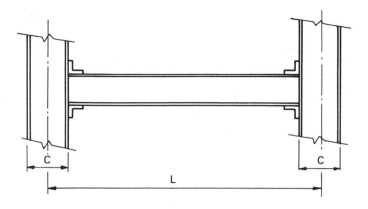

(a) STEEL BEAM ON COLUMNS

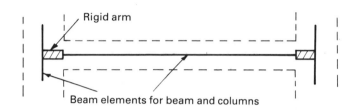

(b) MODEL TAKING ACCOUNT OF THE
    FINITE WIDTH OF THE COLUMN-MODEL A

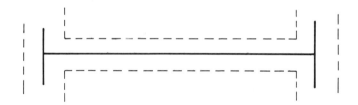

(c) FINITE WIDTH OF COLUMN NEGLECTED - MODEL B

(d) BEAM IN REVERSE BENDING

Fig. 4.3 — Effect of column width. (*Continued next page.*)

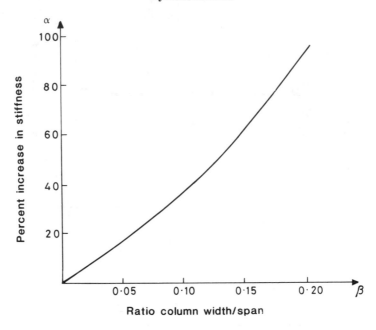

(e) EFFECT OF FINITE WIDTH ON ROTATIONAL STIFFNESS

Fig. 4.3 (*Cont.*) — Effect of column width.

where L is the centre to centre distance between the columns, C is the column width, $\beta = C/L$.

Therefore the percent increase in rotational stiffness when the finite width of the column is considered will be

$$\alpha = \frac{1}{(1 - \beta)^3} \times 100$$

Fig. 4.3(e) shows α plotted against β. This diagram shows that the rigid connection has a significant effect on the beam beam stiffness and can be used as a guide to its significance.

The percentage difference in stiffness under a transverse load on the beam between model A and model B is also equal to α.

The same arguments are valid for the finite depth of the beam in relation to the column stiffness. Fig. 4.10 illustrates various finite width situations for wall models.

4.2.2 Members with non-uniform section
When the members of a frame do not have uniform section, e.g. tapered beams, they can be treated by one of the following methods:

● *By using a beam element with non-uniform cross-section.* These are normally type B elements which, in principle, require mesh refinement. However, a single element with a central node can give fairly good accuracy without refinement.

- *By dividing the member into a set of uniform beam elements.* This effectively uses the uniform element as a type B element. The cross-section at the mid-length of each element can be used. Again, a fine division is not needed.

4.2.3   Shear deformation

When should one include shear deformation? If shear deformation is included in the beam element being used then one might as well give sensible values. High values of shear area will neglect shear deformation.

   In general even if shear deformation affects the overall displacements of a frame significantly it tends not to affect the member forces to the same degree. In conventional frames shear deformation can normally be neglected. In low rise shear walls it can be a dominant factor.

   For calculation of beam deflection, shear deformation becomes more prominent as the span-to-depth ratio decreases. It can be important in I beams since the flanges significantly reduce the bending stiffness but only marginally reduce the shear deformation. An estimate of the effect of shear deformation can be made using the formula given in Table A8.

4.2.4   Torsion

See section 4.4.1.

4.2.5   Connection flexibility

Pin connections for beam elements are a standard feature in analysis packages. It is relatively easy to insert a joint element to represent the connection between structural steelwork members. It is less easy to define the stiffness parameters of the joint.

   For normal purposes full connection fixity will be acceptable for:

- welded joints or bolted joints with web and flange cleats;
- continuous concrete frame construction.

Where smaller sections connect to larger webs, stiffeners are needed to eliminate local rotations. This is also the case where a smaller rolled hollow section is welded to the face of a larger section of this type.

4.2.6   Infilled frames

Brickwork or blockwork which 'infills' the area between beams and columns in a frame makes a major contribution to behaviour. Under lateral load the normal model for the infill is a diagonal compression strut. Stafford Smith & Riddington (1978) recommend that the area of such a strut should be equal to one tenth of the diagonal length of the panel times the wall thickness.

4.3   WALLS

When the analysis of shear walls was suggested to me as a suitable PhD topic in 1963 I could count on one hand the number of published papers on that subject. Ten years later I received a copy of a bibliography on shear walls with over a thousand entries in it. I was therefore fortunate to start working early in an area that has become one of major importance.

4.3.1   Walls without openings
One should treat a wall as a beam element if practicable. Situations where this might not be the case are:

● Where the aspect ratio (height to width) is low.
● Where the lateral spread of vertical load needs to be investigated.

Use of plane stress elements is the viable alternative when beam elements are unsuitable.

When modelling a tall wall one can apply lateral load to either side without affecting the results significantly, but with a low rise wall treated as plane stress elements a lateral point load will cause two components of deformation. First there will be deformation due to localized stress in the area of the load, and second there will be overall deformation due to shear and bending.

Case study 4.2 — Low rise wall
Fig. 4.4(a) shows a low rise wall with height to width ratio = 1.0. It was modelled as

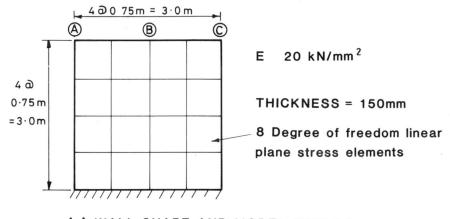

(a) WALL SHAPE AND MODEL DETAILS

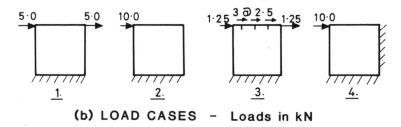

(b) LOAD CASES   –   Loads in kN

Fig. 4.4 — Low rise wall.

sixteen linear strain plane stress elements with the four load cases shown in Fig. 4.4(b). Table 4.1 shows the lateral deflections at points A, B and C at the top of the

Table 4.1 — Deflection of low rise wall (see Fig. 4.4)

Model	Load case	Deflections in x-direction $m \times 10^{-6}$ at points		
		A	B	C
Plane	1	25.6	20.6	25.6
stress	2	32.0	20.6	19.1
	3	22.5	21.6	22.5
	4	16.8	4.7	0.0
Bending and shear deformation		22.9	22.9	22.9

wall. Note the relatively high lateral movement of point A in run 2 which has a single point load at A. This is due to the local deformation characterized by the deformations of load case 4 where the right-hand side of the wall is restrained. Such effects are always present in areas of localized loading but they tend only to be significant where the structure is generally stiff. Note that the bending plus shear model gives a fairly good estimate of the deflection. The plane stress elements are conforming and therefore the plane stress deformations quoted in Table 4.1 will be generally an underestimate.

4.3.2   Frame models for shear walls with openings
In a majority of cases a model comprising beam elements with bending, axial and shear deformation gives satisfactory results for shear wall analysis (MacLeod 1967). Use of plane stress elements for this purpose normally violates Principle 1 stated in section 1.2.

Fig. 4.5(b) shows a frame model of a shear wall. The finite widths of the 'columns' are modelled by rigid arms as described in section 4.2.1. These simulate the bending theory assumption that plane sections remain plane. They are introduced using constraint techniques as described in section 8.1.

The main difficulty in appreciating the validity of a frame model of a wall such as that shown in Fig. 4.5(b) is that a storey height part of the wall normally has a length to depth ratio for which bending theory is not valid — Fig. 4.5(c). Columns are treated as storey height members and designers are accustomed to dividing frames into storey height segments. A wall, however, should be viewed as a building height unit. The full height wall sections on either side of the row of openings of the wall of Fig. 4.5(a) are clearly capable of being modelled as beam elements and the actions from the connecting beams can be treated as loading on such elements — Fig. 4.5(d).

Finite beam depth
The finite depths of the beams or the finite sizes of the columns and the beams can be also be modelled using rigid arms as shown in Fig. 4.6. The model of Fig. 4.6 will overestimate the axial stiffnesses of the columns since they will be axially rigid over

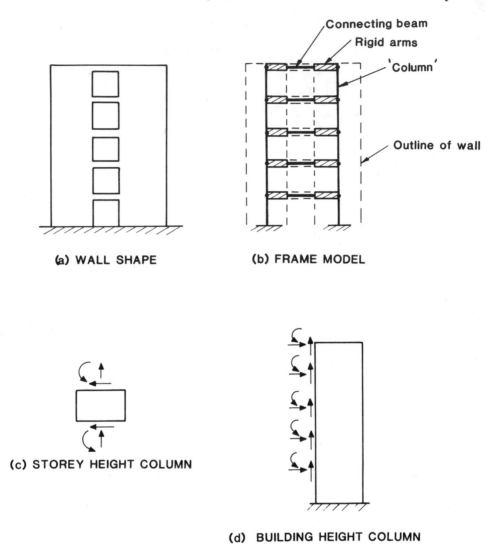

Fig. 4.5 — Frame model for a shear wall with a row of openings.

the depth of the beams. That is, the axial stiffness will be EA_c/h_1 where A_c is the column area, and h_1 is the distance between the rigid parts — (Fig. 4.6). In reality the 'rigid' part will have significant vertical flexibility. This can have a noticeable effect on the overall stiffness of tall walls. This effect can be alleviated by adjusting the area of the columns to give an improved axial flexibility. It is more realistic (though erring on the flexible side) to use an axial flexibility of EA_c/h where h is the storey height. Therefore an equivalent column area of

$$A'_e = \frac{A_c h_1}{h} \tag{4.2}$$

can be used.

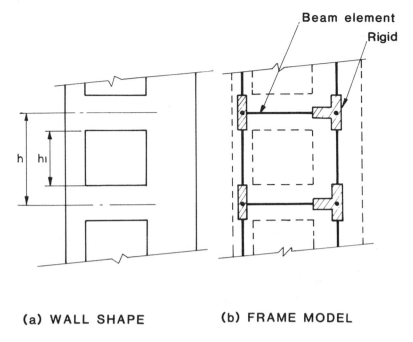

(a) WALL SHAPE   **(b) FRAME MODEL**

Fig. 4.6 — Deep beam frame model.

If the rigid part is modelled using high but finite stiffness then the area of the 'rigid' part can be given a realistic value such as that of the column as suggested for equation (4.2).

Stiffness of connecting beams

The connecting beams in the frame model consist of a lintel beam (if present) and an effective width of floor slab. A common section is therefore a tee beam — see Fig. 4.7.

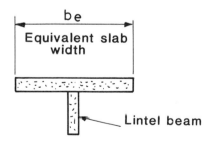

Fig. 4.7 — Section through connecting beam.

The effective stiffness of the slab depends on the following parameters:

• The width of the opening.
• The distance to the adjacent slab support.

- The thickness of the wall.
- The degree of cracking in the concrete.
- The percentage of steel in the concrete.

Rules for selecting equivalent slab widths which include the first three of the above parameters have been formulated (Quadeer & Stafford Smith 1969, Coull & Wong 1981). An accurate prediction of connecting beam stiffness is not presently feasible and the following simplified rule for the effective width of the slab (b_e) may be acceptable:

b_e = lesser value of: opening width, or

0.25 × spacing of walls on each side of wall being considered.

There must of course be adequate shear connection between the slab and the lintel beam for them to be treated as a composite section.

Another factor which received attention was the local deformation at the interface of the lintel beam and the walls (Michael 1967, Bhatt 1973). Some cases where this appeared to be important have been identified but the effect of the floor slab (which is normally present) is likely to dominate the behaviour and the local wall deformation can normally be ignored.

Flange effect of walls

Fig. 4.8 shows a shear wall with a 'flange' wall at right angles to it. The left-hand column of the equivalent frame is treated as a tee section with axis at the centroid of area thus extending the lengths of the rigid arms in this case.

In some cases the flange wall may be quite wide and an equivalent width may need to be defined. If the flange wall has openings a spatial wall approach may be used as described in section 4.3.4.

Column supported walls

Where the wall is supported by columns at ground floor level the basic mode of stress transfer is altered. Vertical stress is transmitted to the columns by an arch action and the transfer beam at first floor level acts mainly as a tie to the arch (Green 1972, MacLeod & Green 1973) — Fig. 4.9. A frame model will not pick up the details of the arch action, and if a more detailed investigation of stress in this area is required, a plane stress element model can be used, e.g. as illustrated in Fig. 4.11.

Case study 4.3 — Shear wall with non-uniform openings

Fig. 4.10(a) shows a reinforced concrete wall with a complex system of openings. Fig. 4.10(b) shows a suitable frame model. In general vertical elements line up to form columns of reasonable length-to-depth ratio and this model should give useful results except as follows:

- At the top left hand corner a single beam element — member A — of low length-to-depth ratio is used to model a column. This would give a poor representation of stress in this area but fortunately the stressses there will be low and therefore the rough model may be acceptable.
- At the bottom right corner of the wall a single beam element — member B — is used to model an area with a fairly complex stress distribution. Much load

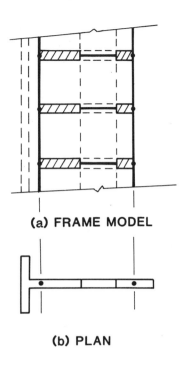

(a) FRAME MODEL

(b) PLAN

Fig. 4.8 — Flange effect of walls at right angles to shear wall.

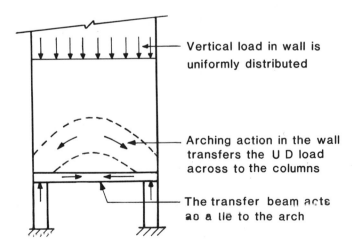

Vertical load in wall is uniformly distributed

Arching action in the wall transfers the U D load across to the columns

The transfer beam acts ao a lie to the arch

Fig. 4.9 — Column supported wall.

redistribution takes place in this area, the element length-to-depth ratio is low, and the stresses are relatively high. A plane stress element model of this area might be worthwhile as described in section 4.3.3.

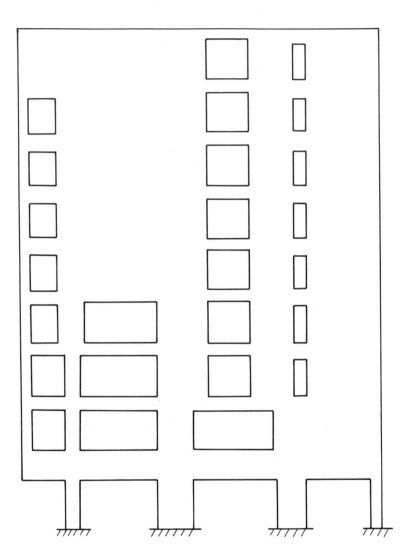

(a) WALL WITH IRREGULAR PATTERN OF OPENINGS

Fig. 4.10 — Frame model for wall with non-regular openings. (*Cont. next page.*)

4.3.3   Plane stress elements
To model a high uniform wall using plane stress elements would normally violate Principle 1 of section 1.2 since the frame model gives similar results using a fraction of

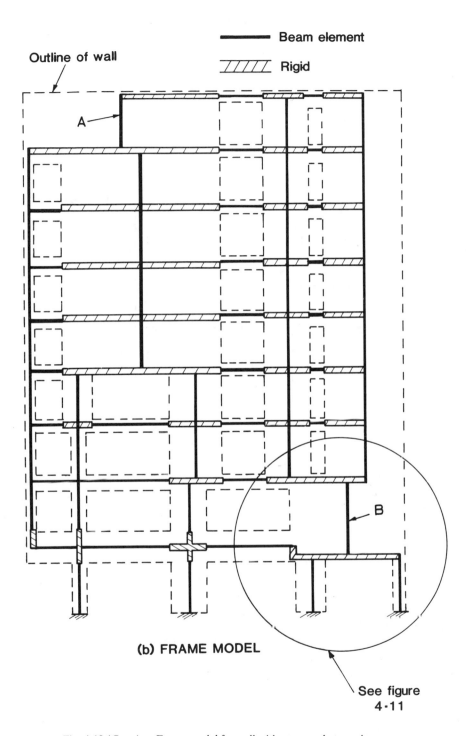

Fig. 4.10 (*Cont.*) — Frame model for wall with non-regular openings.

the number of freedoms. However, such situations as detailed stress in the vicinity of an opening or the redistribution of load on a column supported wall (as noted in the previous section) can be usefully studied using such elements. Fig. 4.11 shows part of

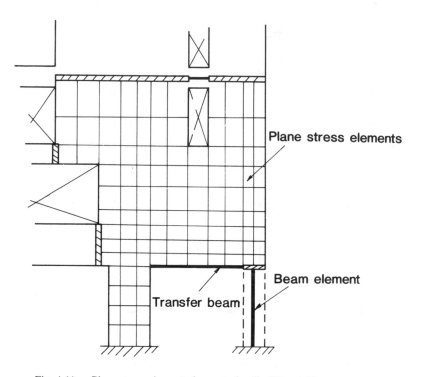

Fig. 4.11 — Plane stress elements for part of wall of Fig. 4.10.

the base of the wall of Fig. 4.10 modelled using plane stress elements and beam elements. At the interfaces with the rest of the wall either forces or deformations from the results of the frame model (Fig. 4.10) are imposed. The relatively slender column is treated as a beam element as is the transfer beam. By modelling the transfer beam in this way one obtains moments and shears which can be used directly in reinforcement calculations.

Reinforcement calculations for the wall modelled as plane stress elements can be based on the rules given in Clark (1976).

4.3.4   Spatial wall systems
Wall systems are often interconnected to a high degree and therefore highly indeterminate. Openings provide further difficulties and it may be impracticable to use a model which predicts detailed behaviour.

Space frame models
Fig. 4.12(a) shows a simple spatial wall system. This is considered to be comprised of three 'walls' whose centre lines are at nodes 1, 3 and 6.

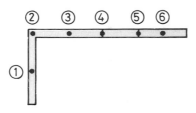

(a) PLAN OF WALL SYSTEM

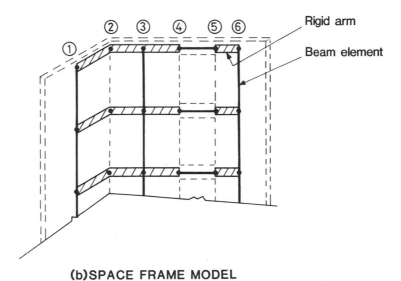

(b)SPACE FRAME MODEL

Fig. 4.12 — Space frame model of wall system.

One option is to treat each wall as a three dimensional beam element and to connect these elements together at each floor level by means of rigid arms or by connecting beams. Fig. 4.12 shows a space frame model of this type. For such models the following factors should be considered:

- This is the same model as used in warping torsion (section 2.2.9) except that the connections between the walls are only at the floor levels rather than continuously with height.
- In such cases it is probably easiest to treat the rigid parts using high but finite rigidity rather than using constraint equations. A method of setting up this model using special plane frame elements with rigid ends is described in MacLeod and Hosny (1977).
- Where a wall is relatively slender and connects to a wall at right angles to it, as illustrated in the core system of Fig. 4.13, the deflection of the model in the plane of the flexible wall (wall c c) has a superimposed double curvature component as illustrated in Fig. 4.13(c). This causes an overprediction of moment in the flexible

column and reduces the stiffening effect of the adjacent connecting beams
(MacLeod & Hosny 1977). This effect can be reduced by adding extra nodes
within the height of the slender wall and connecting them to adjacent extra corner
nodes. This will only be necessary at positions of high stress at the base.
- The effect of the floor slabs in preventing distortion of the cross-section of the wall
  system needs to be modelled. This can be done by either

    (a) assigning relatively high value to the bending stiffnessses of the rigid arms and
        connecting beams in the planes of the floors;
    (b) adding bar elements in the plane of the floors as illustrated in Fig. 4.13(a).

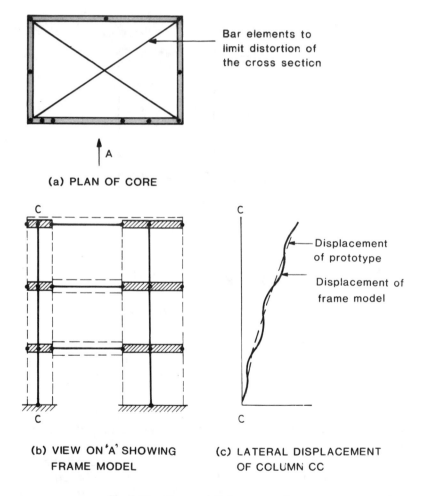

Fig. 4.13 — Core modelled as a space frame.

Warping torsion elements
Another possibility for taking account of warping of the cross section of a wall system
is to treat the complete wall system as a beam element which includes warping torsion
(section 3.4.3). Such models are described in Smith & Taranath (1972) and
Heidebrecht & Swift (1971).

Shell elements

Many spatial wall systems are highly interconnected with irregular openings causing frame models to be unrealistic. Under these circumstances the use of flat shell elements may be feasible but would normally be impractical due to the very large number of elements which would be required. Large order models for buildings are only justifiable where the integrity of the structure is of special importance.

4.4 PLATES IN BENDING AND GRILLAGE SYSTEMS

4.4.1 Grillages

We refer here to true grillage systems formed from interconnected beams and not to grillage models of slabs which are discussed in section 4.4.3.

Grillages are modelled using grillage elements (section 3.4.4) the main problems being to define the torsional stiffness.

Steel construction

If open sections (e.g I sections) are used, warping is an important component of torsion, but since torsional stiffness is low, the normal approach of including only St Venant torsion may be justified.

With hollow sections the torsional stiffness is more important but fortunately the warping component of torsion is less prominent and is normally neglected.

Reinforced concrete construction

With reinforced concrete construction warping torsion is less important. A torsion-less grillage might be used with more confidence but this could result in cracking due to torsional moments which are not provided for in the reinforcement detailing. It is therefore good practice to include the torsional stiffness.

A comprehensive study of torsion in concrete members is given in Tamberg & Mikluchin 1973.

4.4.2 Plates in biaxial bending

A model which includes biaxial bending is needed for two-way spans and one-way spans with non-uniform loading. Two-way spans have a significant structural advantage because their reserve strength is normally high. This is because, as deformations become larger, a membrane type of resistance can develop which is not possible with one-way spans unless the supports are anchored back. Plate bending or flat shell elements or, for concrete slabs, grillage models can be used. These models produce 'lower bound' solutions, i.e they provide moment fields in equilibrium with the applied load to which the lower bound theorem — see Section 8.3.1 — can be applied.

4.4.3 Reinforced concrete slabs

With reinforced concrete slabs use of biaxial bending models is preferable to the Hillerborg strip approach (see Cope & Clark 1984) because they give a moment field which approaches that likely to occur in the real slab. A Hillerborg distribution may not relate closely to the real situation and, although this should result in a satisfactory design for strength, serviceability considerations are not involved. Biaxial bending models should give results which help to reduce cracking.

The output from a biaxial bending model is in the form of bending moments M_x, M_y and M_{xy} — see section 2.2.5 and 8.7.5. For selecting reinforcement these are converted to design moments M_x^* and M_y^* using Wood's Rules (Wood 1969).

Grillage models for concrete slabs

For this type of model the slab is divided into grillage elements. The centroidal axis of an element does not need to coincide with the centroid of the area (Whittle 1985). For edge elements the element axis can be at a distance $d/3$ from the edge where d is the slab depth (Hambly 1976) — see Fig. 4.14.

The second moment of area and shear area of the element is that of the normal cross-section but the torsion constant has half the value of the normal cross-sectional value because of the change in shear flow in the continuous slab as compared with that of a real section of finite size (Hambly 1976). Typical element properties are given in Fig. 4.14.

Uniform transverse loading can be distributed to the grillage members on the basis of the conventional trapezoidal distribution for a rectangular slab but Whittle (1985) suggests that uniformly distributed loads on the grillage members based on a distribution in proportion to the panel dimensions give adequate accuracy. That is, if the side lengths of a panel taking a total uniformly distributed load W (kn) are L_1 and L_2 (m) then the adjacent beams take a U.D. load of $W/2/(L_1 + L_2)$ kN/m.

Allowance can be made for cracking to estimate service load deflection — see (Whittle 1985).

For flat slab construction in buildings, the layout of elements shown in Fig. 4.15 is recommended (Whittle 1985).

With ribbed or beam and slab systems the choice of elements takes account of the variations in cross-section. For example, a support beam can be treated as one of the grillage elements as illustrated in Fig. 4.14. A rib with associated slab can be treated as an element or a group of ribs can be so treated.

The torsional resistance of the elements can be neglected (Fernando & Kemp 1978) since a lower bound solution is still achieved but one risks serviceability problems in doing this. If torsion is included in the model, account *must* be taken of this in the reinforcement detailing.

Case study 4.4 — Flat plate slab using grillage models

Fig. 4.16(a) shows a symmetrical quarter of an interior panel of a flat slab system. Fig. 4.16(b) shows six grillage models used for comparison. The quarter plate is supported vertically at A and rotational restraints are imposed on the four sides about axes parallel to the sides. A transverse uniformly distributed load is applied. Table 4.2 gives the results. The 'Timoshenko' solution (Timoshenko & Woinowsky-Kreiger 1959) does not take account of the finites size of the column and therefore, due to the singularity in the solution at the column support, no finite value of moment is predicted at the column. The 'target' values are those for mesh BE9 from Table 4.3. Note that in the grillage models, the finite size of the column is not taken into

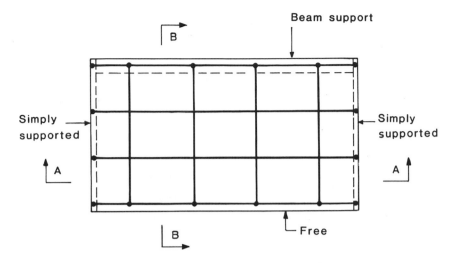

(a) PLAN OF CONCRETE SLAB

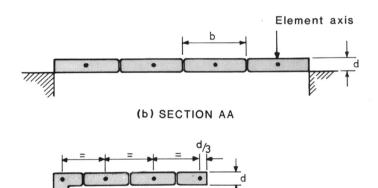

(b) SECTION AA

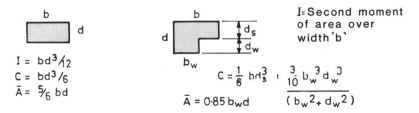

(c) SECTION B-B

$$I = bd^3/12$$
$$C = bd^3/6$$
$$\bar{A} = \tfrac{5}{6}\, bd$$

I = Second moment of area over width 'b'

$$C = \frac{1}{6}\, bd_s^3 \; , \; \frac{3}{10}\, \frac{b_w^3\, d_w^3}{(b_w^2 + d_w^2)}$$

$$\bar{A} = 0{\cdot}85\, b_w d$$

(c) SECTION PROPERTIES

Fig. 4.14 — Grillage model for concrete slabs.

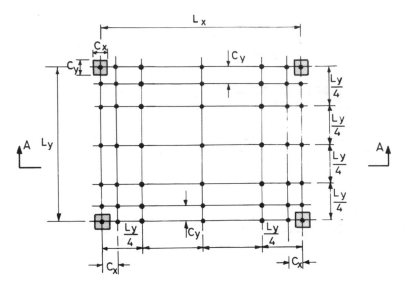

(a) PLAN OF FLAT SLAB PANEL SHOWING GRILLAGE MEMBERS

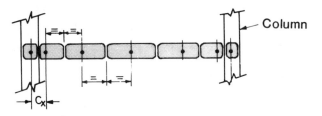

(b) SECTION A–A

Fig. 4.15 — Grillage model for flat slab panel.

account. In Table 4.2 the moment at the column support (M_{XA}, M_{YA}) is quoted at the centreline of the column whereas that for BE9 is the peak value at the corner of the column.

The column moment M_{XA} from BE9 cannot however be treated as a target value since there is still a singularity at the corner of the column.

Model G1 does not give useful results.

The other models generally overestimate M_{XB} and M_{YB} and underestimate M_{XC}. G4, G5 and G6 give sensible values for column moment and the estimates of deflection are reasonable in all cases.

In general the CIRIA mesh, G4, appears to be satisfactory.

Taking account of the finite size of the column was found not to provide better predictions in this case.

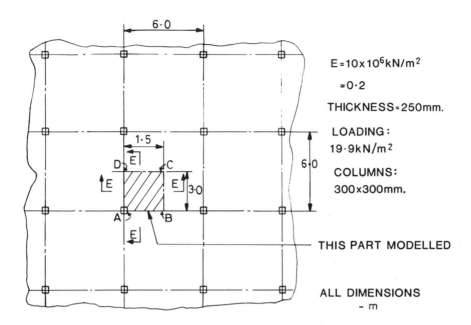

(a) PLAN OF FLAT SLAB SYSTEM SHOWING SYMMETRICAL QUARTER OF A PANEL

$E = 10 \times 10^6 \, kN/m^2$

$= 0.2$

THICKNESS = 250mm.

LOADING:
19.9kN/m²

COLUMNS:
300x300mm.

THIS PART MODELLED

ALL DIMENSIONS
- m

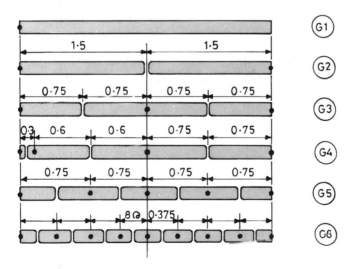

(b) SECTION E-E SHOWING GRILLAGE MODELS

Fig. 4.16 — Flat slab grillage example.

Table 4.2 — Case study 4.3 — grillage models

Model	Vertical displacement at C (mm)	Bending moment (kn m/m)				
		$M_{Xa} = M_{YA}$	M_{XB}	M_{YB}	$M_{XC} = M_{YC}$	
G1	—	− 37.3	+ 22.4	—	—	
G2	10.3	− 76.5	+ 57.9	− 31.7	+ 13.1	
G3	11.4	− 122.5	+ 48.0	− 17.1	+ 15.1	
G4	10.9	− 192.7	+ 45.3	− 18.6	+ 15.7	
G5	11.6	− 164.8	+ 43.5	− 18.9	+ 17.6	
G6	11.6	− 206.4	+ 42.5	− 19.2	+ 18.1	
Timoshenko	11.1	—	+ 35.5	− 13.1	+ 23.6	
Target (BE9)	9.8	− 177.2	+ 35.5	− 12.9	+ 23.1	
BS 8110 Empirical method		− 56.8*	53.8	− 18.9	44.0	

* At face of column.

4.4.4   Plate bending and flat shell element models

Use of plate bending or flat shell elements gives more detailed accuracy than grillage models but is less easy to handle with complex cross-sections.

Choice of element type and mesh

Convergence is discussed in section 3.1.1 and mesh refinement in section 4.1.2. If using quadrilateral elements with corner nodes having three degrees of freedom each (i.e. twelve degrees of freedom elements) a typical mesh for a square panel is 8 elements by 8 elements. The equivalent mesh using elements with mid side nodes would be 4×4. Further refinement of such a mesh might be desirable at corners of openings and column supports, etc.

Ribbed slabs

Ribbed slabs (e.g. waffle slabs) can be modelled using an equivalent thickness. Fig. 4.17 shows a section through a ribbed slab. A pitch length of ribs is chosen, e.g. over the distance b shown in Fig. 4.17. The second moment of area I of the slab cross-section over this width is calculated. The equivalent depth d_e is given by:

$$d_e = \sqrt[3]{\frac{12I}{b}}$$

(4.3)

This thickness models the stiffness of the slab but calculations for stresses in the slab must be based on the real dimensions.

Slabs with different sizes of ribs or different spacings at right angles can be treated as orthotropic slabs using the constitutive relationship defined in equation 2.9.

Values for the terms of this relationship are often based on the following (Timoshenko & Woinowsky-Kreiger 1959):

$$D_x = \frac{EI_x}{1 - v^2} \tag{4.4}$$

$$D_y = \frac{EI_y}{1 - v^2} \tag{4.5}$$

$$D_1 = v\sqrt{(D_x D_y)} \tag{4.6}$$

$$D_{xy} = \frac{1 - v}{2}\sqrt{(D_x D_y)} \tag{4.7}$$

where I_x and I_y are the second moments of area of the slab per unit width and v is Poisson's ratio.

The cross coefficient D_1 and the torsion coefficient D_{xy} as quoted in equation (4.6) and (4.7) are based on the Huber assumptions (Troitsky 1976) and have no theoretical basis.

A slab with variable rib geometry is probably best treated using a grillage model.

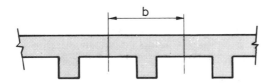

Fig. 4.17 — Section through ribbed slab.

Column supports

Although it is desirable to refine the mesh at a column support since the moment gradients tend to be high at such locations, there is no point in doing this if a point support is used. This type of support (i.e. at a single node) gives a singularity in the solution which for elastic conditions gives a theoretical stress of infinity. It is best to model the finite size of the column as shown in Fig. 4.18. The area of the column is treated as a rigid inclusion with a master node at the column centre-line. Nodes at the column edge are slaves to this master. The master freedoms can be fixed or restrained using springs to model the column stiffnesses.

Methods of imposing the rigid constraints are discussed in section 8.1.

Downstand beams

The centroidal axis of a downstand beam lies below that of the slab and this has an important effect on behaviour. The description which follows is relevant also to upstand beams.

Fig. 4.19(a) and (b) show a concrete system and a composite construction system respectively. A main feature to be modelled is the flange effect of the slab on the

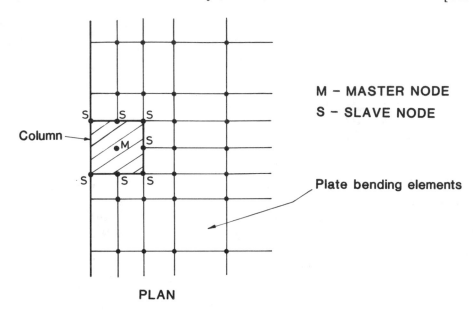

Column

Plate bending elements

PLAN

Fig. 4.18 — Treatment of finite size of column.

beam which results in axial forces being induced in the slab, thus violating one of the main assumptions of the plate bending constitutive relationship. Two ways of modelling this effect are discussed.

Fig. 4.19(c) shows a plate bending and beam element model where the flange effect is modelled using beam elements having an equivalent flange width — Fig. 4.19(d). The rule for doing this given in BS 8110 1985 and is normally satisfactory for this purpose. Plate bending elements are used for the slab part. Such an arrangement keeps the solution size down since there are only three degrees of freedom per node.

A more accurate but more expensive model is shown in Fig. 4.19(e). This uses flat shell elements with six degrees of freedom per node. The beam properties are defined at their centroidal axes and rigid arms (section 8.1) link them to the shell elements as shown. This model is useful where the beams are in reinforced concrete since the member forces from the computer output can be used directly for selecting reinforcement for the beam. The membrane forces in the shell elements must however be considered when selecting the slab reinforcement (Clark 1976).

For composite construction the model of Fig. 4.19(f) is more suitable. The beam element treats only the steel section and the rigid arms have only one dimension since the shell elements continue across the width of the beam — which is not the case for the model of Fig. 4.19(e). Again the membrane forces need to be considered for the reinforcement selection.

Multi-span slabs

If a large area of slab is to be modelled it is likely that only a fairly coarse mesh will be economic for the whole slab. In multi-span situations it may not be worthwhile to model all spans in a single run since interaction between remote spans is low.

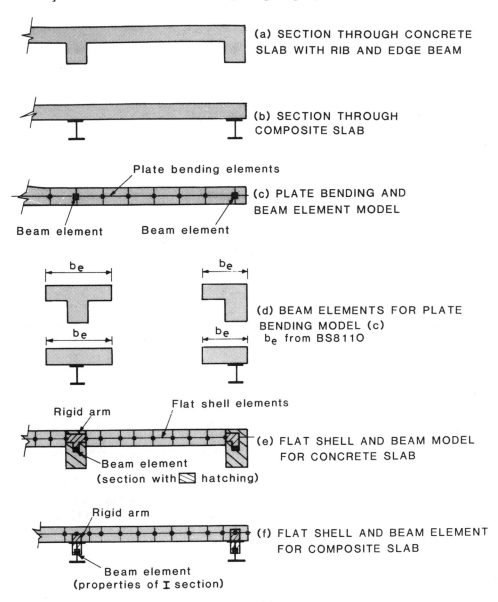

(a) SECTION THROUGH CONCRETE SLAB WITH RIB AND EDGE BEAM

(b) SECTION THROUGH COMPOSITE SLAB

Plate bending elements

(c) PLATE BENDING AND BEAM ELEMENT MODEL

Beam element   Beam element

b_e b_e

(d) BEAM ELEMENTS FOR PLATE BENDING MODEL (c)
b_e from BS8110

b_e b_e

Rigid arm   Flat shell elements

(e) FLAT SHELL AND BEAM MODEL FOR CONCRETE SLAB

Beam element
(section with ⊠ hatching)

Rigid arm

(f) FLAT SHELL AND BEAM ELEMENT FOR COMPOSITE SLAB

Beam element
(properties of **I** section)

Fig. 4.19 — Treatment of downstand beams.

Detailed behaviour should be modelled on smaller portions of the slab with appropriate boundary conditions. These may be based on approximate symmetry conditions or on fixed deformations from a coarser mesh model of a larger part of the slab.

Movement joints

A movement joint in a slab can be simulated using the technique illustrated in Fig. 4.20. Two rows of nodes are created on the line of the joint. They can be on the same

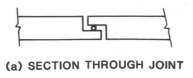

(a) SECTION THROUGH JOINT

**(b) ELEMENT MESH SHOWING LINE
OF DOUBLE NODES**

Fig. 4.20 — Treatment of movement joint.

line if the system allows this or only a very short distance apart if not. Constraints are then imposed to equalize adjacent vertical freedoms at the joint and the adjacent rotations about a horizontal axis at right angles to the joint. The other freedoms remain independent of each other.

Case study 4.5 — Flat slab using plate bending elemennts
The symmetrical quarter of an interior panel of a flat slab system used in Case study 4.4 (page 82) was modelled using plate bending elements. Four noded twelve-degree of freedom hybrid elements were used using the FLASH system (Walder & Green 1981). The boundary conditions were the same as for the grillage models.

Fig. 4.21 indicates the meshes considered. Meshes BE1, BE2 and BE3 do not take account of the finite size of the column whereas the other meshes do this by imposing a constraint over the column area. The Timoshenko results (Timoshenko & Woinowsky-Krieger 1959) do not take account of the finite size of the columns (see Case study 4.4).

Meshes BE8 and BE9 are the most refined and it appears that the predictions for M_{XB}, M_{YB}, M_{XC} and M_{XE} are not significantly affected by mesh division with this degree of refinement. Therefore the BE9 results are taken as target values for these moments.

Note the following points:

- There is a singularity (see Case study 4.4) at A for meshes BE1, BE2 and BE3 and at D for the other meshes, and hence as the mesh is refined higher and higher stresses are predicted at these points.
- The BS8110 values are significantly greater than the bending element values.
- It is not clear what should be taken as the design moment at the column. As the mesh is refined the moment at D increases and that at E decreases. The peak moment at D is very localized and needs to be smoothed out to some degree.

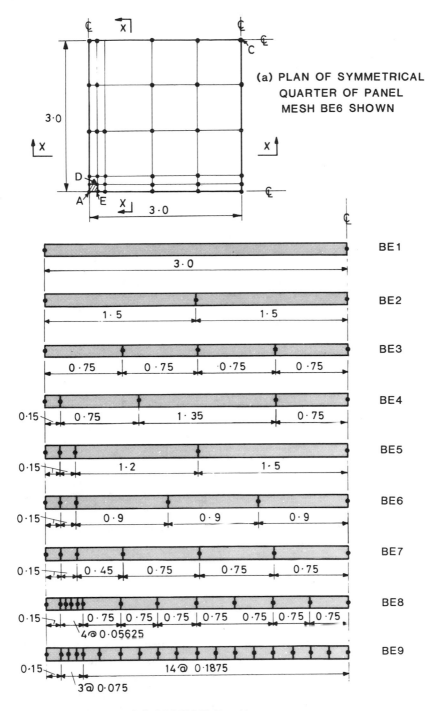

(a) PLAN OF SYMMETRICAL
    QUARTER OF PANEL
    MESH BE6 SHOWN

(b) SECTION X – X

Fig. 4.21 — Plate bending element model of flat slab.

- Meshes BE4, 5, 6 and 7 all give similar results. M_{YB} is somewhat underestimated by BE5. BE4, which is equivalent to the CIRIA mesh (G4 of Table 4.2), appears to give useful results despite the fact that some of the elements are not well proportioned (although the FLASH element used is less senstive to element proportioning than normal).

Table 4.3 — Flat slab plate bending element case study

Model	Mesh for 6 m square panel	Vertical displacement at C (mm)	Beding moment (kn m/m)					
			$M_{XA} = M_{YA}$	M_{XD}	M_{XE}	M_{XB}	M_{YB}	$M_{XC} = M_{YC}$
BE1	2×2	9.9	-95.4	—	—	62.1	-9.6	12.1
BE2	4×4	10.8	-143.8	—	—	39.3	-16.5	25.5
BE3	8×8	11.0	-191.4	—	—	37.6	-13.8	23.8
BE4	8×8	9.4	—	-80.8	-53.1	34.2	-12.5	22.7
BE5	8×8	9.7	—	-147.4	-49.8	39.3	-8.4	22.9
BE6	10×10	9.8	—	-146.1	-48.7	36.2	-11.8	23.2
BE7	12×12	9.8	—	-143.3	-46.8	36.3	-12.0	23.2
BE8	14×14	9.8	—	-197.1	-39.6	35.6	-12.7	23.2
BE9	6×6	9.8	—	-177.2	-39.6	35.5	-12.9	23.1
Timoshenko		11.1	—	—	—	35.5	-13.1	23.6
BS 8110 Empirical method		—	—	—	-56.8	53.8	-18.9	44.0

4.5 STRUCTURE–SOIL INTERACTION

4.5.1 Use of structure–soil interaction models

Structural engineers tend to model superstructures assuming there is no foundation flexibility, whereas foundation engineers treat foundations as if there is no structural stiffness. Foundation movement is often an important factor in structural design and more realistic modelling than this would be worthwhile.

An attractive concept is to use a combined model of the structure and the soil, but such an approach poses significant difficulties. Some factors to be considered are:

- Settlement normally occurs in the long-term and hence non-linear time-dependent soil behaviour should be modelled. Available software to do this is expensive to use and tends to give predictions which do not have a high degree of reliability.
- Models for stiffness of structures likewise may need to take account of long-term behaviour. If the main structural material is steel then long-term effects will not be prominent, but for concrete the effects of shrinkage, cracking and creep can be much more significant than elastic deformations. Masonry behaviour is probably less affected by time-dependent factors but they are not negligible in long-term performance.
- In buildings, the effects of non-structural elements such as partitions and cladding has an important effect on the real stiffness.

- Damage due to settlement is normally caused by differential settlement. The main factor causing differential settlement is the variation of soil stiffness over the plan area of the structure. If this is so, then this variation should be a main parameter of the soil model. It is not normally included in structure–soil interaction models.

One tends to assume that the soil presents a more difficult modelling problem than the structure. However, in settlement situations, there is no merit in making a non-linear time-dependent model of the soil unless the structure is modelled to an equivalent degree of accuracy. To achieve this is probably not a practical proposition at present for most structural systems.

Differential settlements have been predicted to reasonable accuracy but such outcomes tend to be based on extrapolation from observed behaviour rather than on predictive modelling.

Having come to this conclusion, we ask the question 'Is structure–soil interaction modelling worthwhile if it cannot be predictive?'. This is an important question, the essence of which is not confined to structure–soil interaction.

When an analytical model does not give good prediction of absolute behaviour, I like to assume that it can provide more reliable information about relative behaviour. For example, if one carries out a parameter study to investigate the effect of a particular assumption in a model, then the conclusion drawn may have a validity even although the prediction of absolute behaviour is unreliable.

Structure–soil interaction problems provide us with good examples of how such a principle can be applied. We set up an elastic model which includes the soil and the structure. This will not give any useful information about what the amount of settlement will be but could, for example, indicate that the soil movement will not affect the structure significantly or that stiffening the beams will (or will not) reduce differential settlement.

Acceptance of such conclusions from a crude model can be an act of faith but such acts are not uncommon in structural design.

The structure–soil interaction models discussed in this section therefore can be used for:

(1) Assessing the relative importance of components of the model for settlement situations.
(2) Providing a set of internal forces in a structure in equilibrium with the applied load which can be used for strength calculations justified by the lower bound theorem (Section 8.3.1).

Institution of Structural Engineers (1989) is a useful reference on this subject.

4.5.2 Foundation models

Pad footings for columns

Such footings are normally treated as rigid blocks, as shown in Fig. 4.22. In some cases it may not be necessary to model the finite depth of the footing.

Foundation beams

These would be treated as beam elements with the finite depth taken into account if desired, as shown in Fig. 4.23.

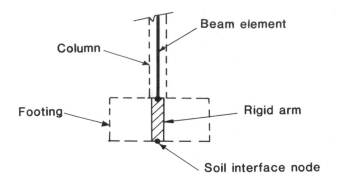

Fig. 4.22 — Treatment of a footing.

Rafts

Rafts can be treated using any of the models described in section 4.4.

For deep rafts it may be worthwhile to take account of shear deformation, e.g. by using a plate bending element based on the Mindlin constitutive relationship (see section 3.6.2) or by including shear deformation in a grillage model.

Cellular rafts can be treated using an equivalent thickness using the relationship of equation (4.3). Alternatively the detailed behaviour of a cellular raft can be modelled using flat shell elements.

If a raft is analysed separately from the structure the stiffening effect of a wall on the raft can be modelled by imposing rigid constraints (section 8.1) along the line of the base of the wall.

Piles

The stiffness would normally be characterized by springs at the top of the pile. The choice of spring stiffness depends on several parameters including

- pile geometry
- E value for pile
- soil characteristics
- whether or not the pile is end bearing
- the interaction of piles within a group.

This is a complex problem — see Poulos & Davis (1980).

Interface with soil

It is normal to neglect the horizontal friction between the base of the foundation and the soil. This is likely to be a conservative assumption.

4.5.3 Soil models

Winkler model

This is discussed in sections 2.5.1 and 3.9.1. Beam or plate bending elements with Winkler support terms make introducing the soil stiffness an easy matter but

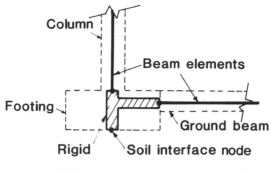

(a) BEAM WITH FOOTING

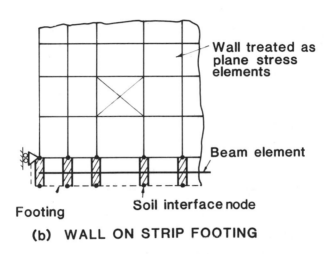

(b) WALL ON STRIP FOOTING

Fig. 4.23 — Treatment of foundation beams.

springs at the soil interface nodes give similar results as far as overall behaviour is concerned.

Typical pressure distributions under a rigid base are illustrated in Fig. 4.24. The effect of the shear transfer with cohesive soils is to increase the pressure at the edges of the foundation. The reduction of pressure found at the edges of a rigid base with cohesionless soils is probably due to local failure in this region. The 'cohesionless' pressure distribution shown in Fig. 4.24 may not occur with deeper foundations since confining pressure will induce shear transfer.

The Winkler model may therefore be more suitable for cohesionless soils but gives a poor representation of the pressure distribution with cohesive soils. This can cause severe discrepancies in the prediction of moments in comparison with more accurate analysis (Hemsley 1988).

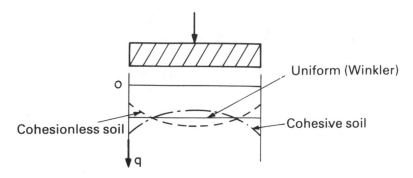

Fig. 4.24 — Soil treatment below a rigid base.

Elastic half space
This is discussed in sections 2.5.2 and 3.9.2 (see also Wood 1977). Elastic half space elements are not commonly available in standard packages but are preferable to the Winkler model.

Plane strain or solid elements
Plane strain or solid elements can be used with a wide variety of constitutive relationships. The computer time required for runs using solid elements with non-linear material properties is lengthy and specialized knowledge is needed. Such models may however soon become more widely used.

4.5.4   Structure models for structure-soil interaction
One can use a 3D model of the structure (as discussed in section 4.7.1 for example) but some alternatives are discussed here.

Interface structure matrix
One can devise a model of the structure defined at the soil interface by applying successive unit vertical movements at each support of a 3D model of the structure in turn with the other supports fixed vertically. The support reactions from one of these load cases provides one column of a structure stiffness matrix. The complete stiffness matrix can be set up in this way and added to a stiffness matrix which represents the soil. This might be a useful approach if the structure is to remain elastic but where the soil behaviour is to be given more detailed treatment.

Equivalent raft
For multi-storey buildings an equivalent raft can be used to simulate the stiffness of the structure (Hooper and Wood 1976). Walls and wall systems can be treated as fully rigid over their area and the raft second moment of area based on the sum of the floor slab stiffnesses.

Equivalent frame
In one of the earliest papers on structure–soil interaction Meyerhof (1947) modelled a frame without diagonal bracing as an equivalent shear beam. The concept he used

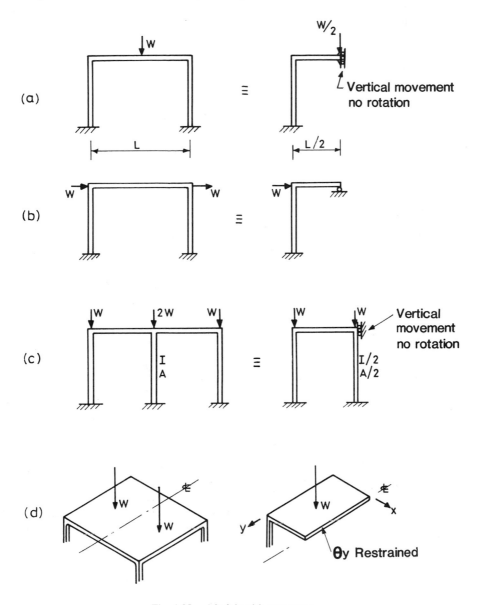

Fig. 4.25 — Models with symmetry.

had an error in it and a beam model based on the recommendations given in section 6.4 would give better results.

4.6 USE OF SYMMETRY CONDITIONS

To reduce the size of analytical models it is often convenient to take account of symmetry conditions (Glockner 1973). The main problem is to define the boundary restraints at the axes of symmetry. This is done by identifying the movements which are inhibited by symmetry on the axes of symmetry. The rule is — If for a given degree of freedom on an axis of symmetry the deformation is zero under symmetry conditions then that deformation should be restrained. If the deformation is possible under the symmetry conditions then it is not restrained. Members which lie on axes of symmetry have their sectional properties halved. Fig. 4.25 shows some typical symmetry restraints.

If the structure is symmetrical then any loading system can be treated as a combination of symmetrical cases. Fig. 4.26 shows a symmetrical portal frame with

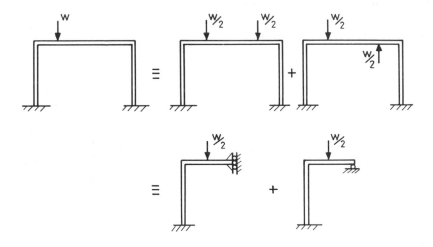

Fig. 4.26 — Treatment of general loading using symmetry.

an unsymmetrically placed load. This can be treated as the sum of symmetric and skew-symmetric cases as shown in the diagram. The boundary conditions are different in each case, so two quite separate computer runs are needed. Therefore the size of the structure to be analysed is reduced by half but combining the results from the two runs may require some extra effort on the part of the user.

Fig. 4.27 shows a pyramid system which has diagonal axes of symmetry. It is possible to model this using a symmetrical eighth provided the program used allows restraints to be imposed on axes which are not parallel to the co-ordinate axes.

4.7   MODELS FOR BUILDING STRUCTURES

4.7.1   Three dimensional models for multistorey buildings

Fig. 4.28 is a plan of a somewhat idealized four-storey building. This building shape is used to demonstrate some modelling techniques.

Fig. 4.29 shows some of the features of a 3D model of the building of Fig. 4.28.

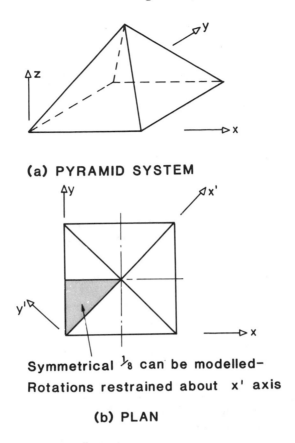

(a) PYRAMID SYSTEM

**Symmetrical ⅛ can be modelled–
Rotations restrained about x' axis**

(b) PLAN

Fig. 4.27 — Treatment of symmetry for system with diagonal axes.

Disadvantages of a 3D model are that the data required may be large and the computer run time may be long. Advantages are that decisions regarding distribution of lateral and vertical load are not required. Vertical load take-down is done automatically and the process of conceiving a model is easier than when a set of plane frames are used.

Vertical supports
The columns are treated as vertical beam elements with properties as discussed in sections 3.4 and 4.2.

Walls and wall systems (e.g. cores) are also treated as vertical line elements but their finite size in plan cannot be neglected. The normal way to take account of this finite size is to neglect warping of the cross-section and to define a rigid plane to which the beams are connected. This is an extension of the method of taking account of the finite size of columns discussed in section 4.4.4.

In Fig. 4.29 a constraint equation approach is illustrated. A master node is defined on the centroidal axis of the section and slave nodes can be introduced at any

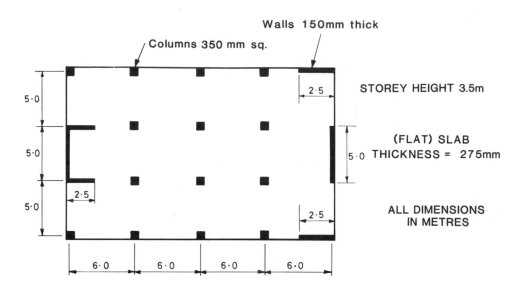

Fig. 4.28 — Plan of 4-storey building.

point on the cross-section which requires a connection to another element. The rigid planes can also be defined using members with high but finite stiffness — see section 8.1.

If warping of the cross-section of the wall system needs to be modelled then a more detailed frame model can be used as discussed in section 4.3.4.

If structure–soil interaction (section 4.5) is not considered it is normal to assume full fixity at the base of vertical supports (Ellis 1980) unless pin supports are specifically provided.

Floor systems

The floor system has three main structural functions:

(1) To distribute the vertical load in bending to the vertical supports.
(2) To distribute the lateral load by in-plane action.
(3) To transmit vertical shear forces between the lateral supports to make them act compositely.

In an overall analysis of a multistorey building one would not normally include a detailed model of the bending action of the floor slabs — see Principle 5 in section 1.2.

Possibly the simplest approach is to use beam elements for any beams that are present and a single flat shell element for each floor slab panel, as in Fig. 4.29. With more irregular layouts it will be necessary to introduce more than one flat shell element per panel as shown in Fig. 4.29. The model shown would not be used for proportioning reinforcement in the floor slabs except that needed to cater for in-plane forces due to lateral load transfer (although such reinforcement is not normally specified).

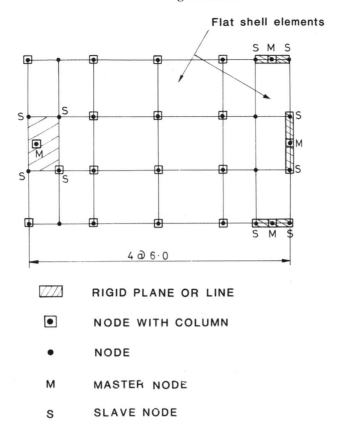

Fig. 4.29 — Plan showing 3D model of multistorey building.

The downstand effect of the beams would normally be ignored at this stage. The thickness used for the shell element is either the actual thickness or an equivalent thickness for ribbed or waffle slabs as described in section 4.4.4.

With such a model, element loads will be distributed to the vertical supports in a sensible proportion and the in-plane stiffness of the floor is realistically treated.

If a shell element is not available in the analysis package then a 'grillage' model as shown in Fig. 4.30 can be used. Fig. 4.30(a) shows such a model with only members on the column lines. It is probably unrealistic to take the half span of the floor as the width of a beam, so taking one quarter of the floor span on either side as the effective width as shown on the cross-section is a commonly used rule.

The I value for the in-plane bending of the beam elements can be calculated on the basis of the cross-section shown. The moment connections between the beam elements in the plane of the slab will provide a fairly stiff diapraghm action for the floor. If extra in-plane stiffness is required one can introduce diagonal elements in the panels. It is probably best to make these elements pin connected in order to avoid introducing extra out-of-plane stiffness to the panels. The vertical loading can be applied to the grillage elements, as discussed in section 4.4.3.

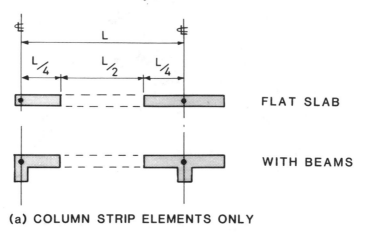

(a) COLUMN STRIP ELEMENTS ONLY

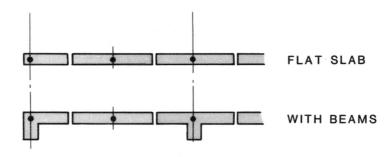

(b) COLUMN STRIP AND MIDDLE STRIP ELEMENTS

Fig. 4.30 — Grillage elements for slabs in 3D models.

Fig. 4.30(b) shows a similar model with the addition of middle strip members. This requires double the number of nodes as compared with that of Fig. 4.30(a) but gives a better representation. With beam and slab systems it might be sensible to take the column strip beams as having an equivalent flange width from the rule in BS 8110 1985 as shown. For flat slab systems the normal rule for column strip and middle strip could be used. The validity of this model for bending action in the slab is discussed in section 4.4.3.

For properties of beams which connect walls to form composite shear wall systems see section 4.3.2.

Foundations
See section 4.5.

4.7.2   Plane frame models for lateral load
The building illustrated in Fig. 4.28 would probably not require special treatment for lateral load. The core and wall at either end would be likely to be more than adequate

for this purpose. More detailed models for this structure are discussed here to illustrate the techniques with no recommendation as to the degree of model refinement needed for this situation.

Interconnected plane frames

Fig. 4.31 shows a plane frame model of the four-storey building of Fig. 4.28 for loading in the y-direction. The lateral load resistant units are drawn separately and then connected at storey levels by rigid links which simulate the in-plane rigidity of the floors. Note the following:

- The connecting links have no bending stiffness and can be created using either bar elements with high axial stiffness or by constraining the lateral freedoms at each storey level to be equal for all frames.
- Columns adjacent to a shear wall such as those of Frame 1 are often neglected in this type of model.
- The three central frames are identical and therefore are treated as a single equivalent frame with member cross-sectional properties having three times the value for one frame. The member actions in the equivalent frame are divided by three to give the estimates for a single internal frame.
- The finite widths of the core and the wall are taken into account as shown.

Equivalent single wall and frame

Fig. 4.31(b) shows a simplified model of that of Fig. 4.31(a). A single wall is defined which has bending stiffness equal to the sum of the separate wall stiffnesses. In this case the I value would be the sum of I values for the core and for the wall part of frame 5.

A single frame is defined which has the following properties:

- The storey heights are the same as that of the prototype.
- At any storey level each column of the frame has an I value of

$$I'_c = \frac{\Sigma I_c}{2}$$

  where I_c is the I value of a column in the prototype and the summation is over all the columns in the building at the given storey level.
- The span of the beams of the frame is arbitrarily fixed at l' metres ($l' = 1.0$ is the obvious choice).
- At any storey level the I value of a beam (I'_b) is found from the relationship:

$$\frac{I'_b}{l'} = \sum \frac{I_b}{l}$$

  where I_b is the I value of a beam in the prototype and l is the span of that beam. The summation is over all beams in the building at the given storey level.

The equivalent frame and the equivalent wall are connected together in the same way as described for the model of Fig. 4.31(a) and the load applied.

The member forces from the equivalent model are distributed to the members of the prototype in the following proportions:

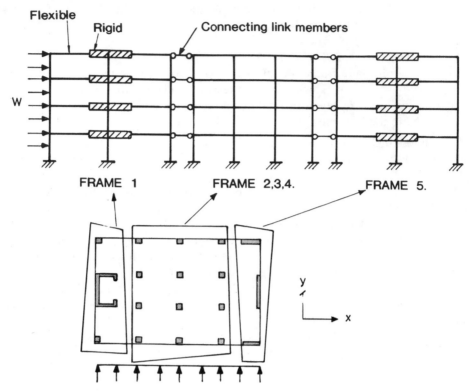

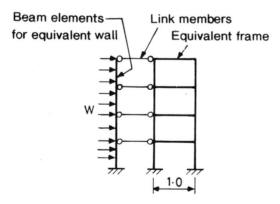

(a) INTERCONNECTED PLANE FRAMES.

(b) EQUIVALENT SINGLE WALL AND FRAME.

Fig. 4.31 — Plane frame models.

- The column moments and shears in the proportion: $I_c/(\Sigma I_c/2)$.
- The beam moments in the proportion: $I_b/l/(\Sigma I_b/l)$. The beam shears are M_b/l where M_b is the end moment in a beam.
- The wall moments and shears in the proportion — $I_w/\Sigma I_w$ where I_w is the I value of a wall of the prototype.

The model of Fig. 4.31(b) may give a fairly crude representation of the protype structure but it can be set up quite quickly and can be used to give a reasonable estimate of the relative importance of the walls and the frames in lateral load transmission.

An approximate solution technique for the system of Fig. 4.31(b) is described in section 6.6.2.

4.7.3   Rigid plate on springs model for lateral load
This model gives important insights for understanding of lateral load behaviour of multistorey building structures. Although the behaviour of a complete building is defined using a small number of degrees of freedom it gives remarkably good accuracy as compared with more complex models.

Fig. 4.32(a) and (b) show the system of Fig. 4.28 divided into frames which are considered to resist lateral load — described as 'support frames'. The basis of this model is that the floor slabs are represented by a rigid plane supported by springs which simulate the effect of the frames.

Fig. 4.32(c) shows how the spring stiffnesses are calculated. A plane frame model of each frame is set up and a lateral load W applied to find a top lateral deflection Δ. The spring stiffness k_i is then given by

$$k_i = \frac{W}{\Delta} \tag{4.8}$$

(Note that if a symmetric half of the frame is used to calculate k_i then $2W$ should replace W in equation (4.8).)

W can be a uniformly distributed load or the total lateral load on the building (which may not be uniform with height).

Fig. 4.33(a) shows a rigid plate on springs model for the example building. This uses constraints (section 8.1) to model the in-plane action. The springs can be treated either as joint elements or as bar elements with axial stiffness k_i.

Fig. 4.33(b) shows the plate treated as beam elements. These need to have sufficient rigidity to link the springs in a way that simulates the in-plane rigidity of the floor slabs.

Fig. 4.33(c) shows a beam element on springs model where the springs in the x direction have been neglected The I value ot the beam is chosen so as to model the in-plane rigidity (i.e. it should be stiff compared with the spring stiffness — but not too stiff).

A load statically equivalent to the total lateral load is applied to appropriate freedoms and a solution obtained. A spring load from these results is denoted as P_i. This is now treated as a load having the same distribution with height as the applied load. Analysis of frame i under this load gives the relevant member actions.

If, when calculating k_i, a computer model of the support frame is used with the total lateral load on the building (W) applied, then the results from this model

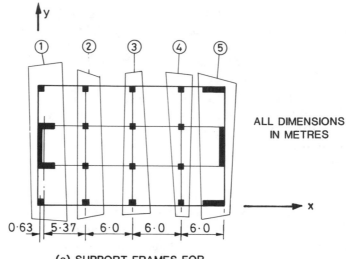

**(a) SUPPORT FRAMES FOR
y DIRECTION SPRINGS**

ALL DIMENSIONS
IN METRES

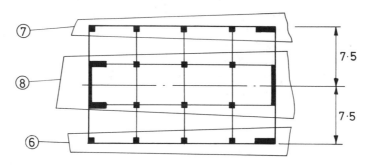

**(b) SUPPORT FRAMES FOR
x DIRECTION SPRINGS**

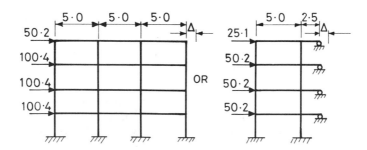

(c) CALCULATION OF TOP STIFFNESS

Fig. 4.32 — Division of building structure into support frames.

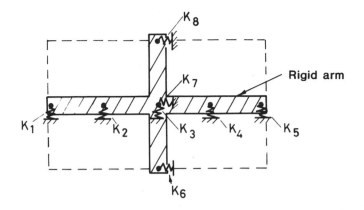

(a) CONSTRAINTS TO MODEL IN−PLANE ACTION

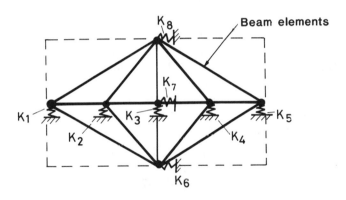

(b) BEAM ELEMENTS TO MODEL IN−PLANE ACTION

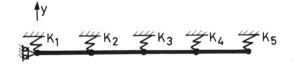

(c) SINGLE BEAM MODEL
(x Direction springs neglected)

Fig. 4.33 — Rigid plate on springs model.

are multiplied by the factor P_i/W to establish the member actions in that support frame.

The basic assumptions for such models are:

- The floor slabs are rigid in their own planes.
- All the lateral load resistant units have the same mode of lateral displacement.

The second of these assumptions is the main source of approximation. Frames tend to deform basically in a shear mode (section 6.4) whereas walls deform in a bending mode — see Fig. 3.7. When these are interconnected there are non-uniform interacting forces and the resultant load on a wall or on a frame will not have the same distribution with height as the load on the building.

These models can be solved by hand calculation as described in section 6.6.

Case study 4.6 — Four-storey building
The analysis under lateral load of the building of Fig. 4.28 is investigated. Four models are considered:

M1: The interconnected plane frame model of Fig. 4.31(a).
M2: The equivalent wall and frame of Fig. 4.31(b).
M3: The rigid plate on springs — as Fig. 4.33(a).
M4: The rigid beam on springs — as Fig. 4.33(c) — see also Fig. 6.10.

The applied lateral load (in the y-direction) was $1.2\,kN/m^2$, giving a total load on the building of $24 \times 14 \times 12 = 403.2\,kN$. A value of $E = 20\,kN/mm^2$ was used and shear deformation of the members was neglected.

The centre line of frame 1 is taken at the centre of area of the core since the core will dominate the behaviour of this frame.

Member properties used are:

Length units: metres. x and y represent the global axes.

Columns	$I_x = I_y = 0.001251$	$A = 0.1225$
Core	$I_x = 7.188$	$I_y = 0.977$
3 m wall	$I_x = 1.56$	$I_y = 0.001406$
2.5 m wall	$I_x = 0.000703$	$I_y = 0.1953$

1.5 m wide edge beam	$I = 0.0026$
3.0 m wide internal beam	$I = 0.0052$
1.25 m wide edge beam	$I = 0.00217$
2.5 m wide internal beam	$I = 0.00433$

For M2 the member properties used were

$$I_c = \frac{\Sigma I_c}{2} = \frac{0.001251 \times 14 + 0.007703 \times 2}{2}$$

$$\Sigma I_b = 9 \times 0.0052 + 4 \times 0.0026 = 0.0572$$

$$\Sigma I_w = 7.188 + 1.56 = 8.748$$

To calculate the spring stiffnesses for models M3 and M4 each support frame was analysed separately under the total load of 403.2 kN. For frame 1, for example, this gave a top deflection $\Delta = 0.934\,mm$. The top stiffness is therefore:

$$k_1 = \frac{W}{\Delta} = \frac{403.2}{0.934 \times 10^{-3}} = 431.6 \times 10^3 \text{ kN/m}$$

The other spring stiffness used were:

$$k_2, k_3, k_4 = 9.93 \times 10^3 \text{ kN/m}$$
$$k_5 = 113.5 \times 10^3 \text{ kN/m}$$
$$k_6, k_7 = 59.82 \times 10^3 \text{ kN/m}$$
$$k_8 = 273.1 \times 10^3 \text{ kN/m}$$

The results of M3 give a spring force for frame 1 of

$$F_1 = 199.2 \text{ kN}$$

The results of the separate analysis of frame 1 is therefore multiplied by the factor:

$$\frac{199.2}{403.2} = 0.496$$

to give the predicted values of internal actions.

The slab inertias were based on an effective width of slab equal to one quarter of the slab span to either side (e.g. for an internal y-direction frame, slab inertia $= 3.0 \times 0.275^3/12 \text{ m}^4$).

For M3 and M4 the in-plane rigidity was modelled using the constraint equation facilities of FLASH (Walder & Green 1981) and bar elements were used for the springs.

Table 4.5 gives the results. For comparison, the lateral movement in the y-direction at the top of frames 1 and 5 and the base moment in the core of frame 1 and in the central wall of frame 5 are considered.

The y-direction stiffness of the core of frame 1 and the corresponding wall in frame 5 dominate the behaviour and hence the columns could be neglected in a lateral resistance calculation. Note the significant increase in load to frame 5 when torsion is considered. The stiff lateral supports are not well proportioned, causing the plane frame approach to overestimate moment in frame 1 and underestimate it in frame 5.

The single wall and frame model, M2, gives reasonable results compared with M1 but the wall moments are higher with M2 because of the frame action in walls 1 and 5 in M1.

The results for M3 and M4 are similar, showing that the x-direction springs have little effect on behaviour under load in the y-direction. The advantage of M3 is that it can also be used for x-direction lateral load.

4.7.4   Single-storey buildings

Fig. 4.34 shows a typical pitched portal frame structure. The normal approach to load distribution is on the basis of a single portal frame with the lateral and vertical load based on frame spacing. This distribution of vertical load will be close to the real situation but the real lateral load distribution is likely to be quite different. What happens with lateral load (in the planes of the portal frames) is that the external

Table 4.5 — Lateral load models of a four-storey building

Model	Figure number	Top displacement in y-direction (mm)		Base moment (kN m)	
		Core of frame 1	Central wall of frame 5	Core of frame 1	Central wall of frame 5
M1	4.31(a)	0.676	0.671	2087	428
M2	4.31(b)	0.735	0.735	2156	468
M3	4.33(a)	0.477	1.484	1381	1014
M4	4.33(c)	0.448	1.587	1296	1085

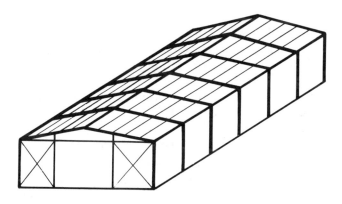

Fig. 4.34 — Pitched portal structure.

frames are much stiffer than the internal ones since they are normally braced and sheeted. The roof sheeting in combination with a wind truss in the plane of the roof causes the lateral load to be transferred outwards to the external frames.

The separate portal frames are normally proportioned for strength using the plastic method (Horne & Morris 1981) and an elastic analysis is used to check stiffness. This procedure is likely to be conservative since the contribution to strength and stiffness of the end frames is significantly underestimated.

An attractive concept is to use a space frame model of the building structure. Note again, however, that in some cases codes of practice are tuned to methods of analysis and one should not adopt unconventional methods without care.

The modelling procedure is straightforward except for the treatment of the sheeting if that is to be included. A standard reference on this subject is Davies & Bryan (1982). They have carried out extensive shear tests on sheeting panels. Their test arrangement is shown in Fig. 4.35. The shear S is applied and δ — the deflection

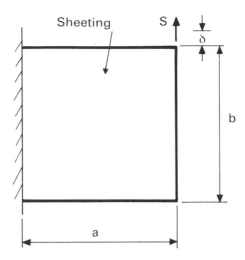

Fig. 4.35 — Shear flexibility of sheeting panel.

in the line of S— is measured. The depth of the panel is b and the length a. (Note that Davies & Bryan use a as the length of the panel perpendicular to the corrugations and b as the length parallel to the corrugations.) Here a and b are defined in relation to the loading direction (b is in the direction of S).

The flexibility of the panel is defined as

$$\delta = cS \tag{4.9}$$

where c is a flexibility factor.

Data for estimating c are given in Davies & Bryan (1982). For lateral load, Davies & Bryan model the system as a set of plane frames connected by springs which have stiffness $1.0/c$. This assumes that the sheeting acts in a pure shear mode (Fig. 3.7(b)) and gives good correlation with test data. There will, however, be a bending mode component where the eaves beams, purlins and sheeting itself will provide 'flanges' for an equivalent beam which spans between the end frames partially supported by the internal portal frames. If such bending action is significant then the shear stiffness of the sheeting can be modelled representing each sheeting panel by a tension diagonal. The area of a diagonal member — A_d — is based on:

$$A_d = \frac{b}{E \sin^3 \theta c} \tag{4.10}$$

where $\theta = \tan^{-1}(b/a)$.

Equation (4.10) is derived in section 6.2.2.

4.8   NON-LINEAR MODELS

4.8.1   Solution methods

The iterative process used in non-linear solutions can use up large amounts of computer time and can diverge from (rather than converge to) the sought result.

Objectives in selecting solution methods are therefore to keep the solution time down and to preserve the stability of the solution.

Like many other situations, simple traditional methods can give good results. Possibly the best known method for non-linear solutions is the Newton–Raphson method. This is commonly used in finite element work and is briefly described here to illustrate a typical method.

Fig. 4.36 illustrates how the Newton–Raphson method proceeds. Starting from a

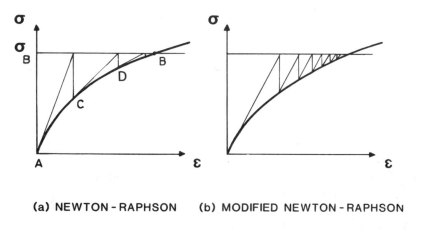

(a) NEWTON - RAPHSON   (b) MODIFIED NEWTON - RAPHSON

Fig. 4.36 — Newton-Raphson solution method.

known position A we want to find the solution at point B on the stress strain diagram. Firstly, the tangent stiffness at A is established. Using this, an *estimate* of strain at stress σ_B represented by the point C is established. A new tangent stiffness at the point C is set up from which a new estimated position at D is established. This procedure is repeated until point B is reached. The disadvantage of this approach is that a new set of equations needs to be set up and solved for each step.

The modified Newton–Raphson method illustrated in Fig. 4.36(b) is also commonly used. The tangent stiffness at A is used at each step in the approach to B. The advantage is that the stiffness matrix needs to be set up only once and each step represents a solution for a different load case for the same matrix. A greater number of steps are needed than with the non-modified method but each step takes much less computer time.

The modified Newton–Raphson approach is sometimes used within a load increment and a new tangent stiffness formulated for the next increment.

4.8.2   Plastic collapse of frames
The traditional method of calculating the plastic collapse load of a steel structure is the limit analysis approach where possible collapse modes are searched to find the one that gives the minimum collapse load. This procedure requires one to assume that the loads are proportionate, i.e. that the vertical load and the lateral load increase together in a fixed ratio. This is not the way that real structures receive load, and since the results with non-linear behaviour are path dependent a better result can be achieved by incrementing the load in a pattern closer to the real situation, e.g. by

applying the dead load and then the live load. Doing this probably will not make a significant difference to the predicted collapse load but it is one less assumption that need be made. The major advantage of the load stepping approach for non-linear modelling of frames is that one can observe the development of plastic hinges and the degradation of stiffness as the load is applied and when the collapse load is obtained one does not need to worry about the possiblility of having missed a collapse mode provided the load steps are sufficiently small.

The most rigorous approach is to successively insert joint elements with elasto-plastic behaviour where plastic hinge positions are identified. It is more common however to introduce the joint elements at the outset with high elastic stiffness and allow them to go plastic as appropriate. In such cases one has to be sure that one does not miss a possible plastic hinge location. Such locations on the span of beams with distributed transverse load may not be easy to identify at the outset. If the program only tests for yield at the joint elements it may be necessary to check that the plastic moment is not exceeded at other parts of the structure.

Another approach is to increment the load and successively insert pin connections where plastic hinges are identified. The plastic moment is then applied as a bending moment load at each pin connection. The results of successive load increments are then added as the load is incremented. A disadvantage of this approach is that plastic hinges may form and then close as the pattern of plastic hinges develops. Such closure of plastic hinges may be relatively uncommon.

Shear wall frames
Plastic hinges can be inserted in the frame model of a shear wall with row openings of the type shown in Fig. 4.5 and a plastic collapse load estimated. The main hinge positions will be at the bases of the columns and at both ends of the connecting beams. Note however that the connecting beams need high ductility if they are all to remain plastic up to collapse (Paulay 1969).

4.8.3 Non-linear geometry effects in frame structures
The elements discussed in section 3.11 are used to take account of the non-linear geometry effects.

It is common to extract a frame from a structural system and to apply vertical load to it to estimate its buckling load. This is probably a safe procedure but can be quite far from the real behaviour if the buckling is in a sway mode. Frames do not exist in isolation and single frames in buildings for example do not just buckle sideways on their own but will be braced by the rest of the structure. For sway buckling the system must go — not just a single frame. It is sensible therefore, wherever practicable, to include all lateral resistance in the model when estimating sway buckling. This concept is discussed in section 6.8.

Design procedures have, however, been established on the basis of bare frame analysis and care needs to be taken with more accurate modelling in relation to code provisions.

4.8.4 Non-linear analysis using flat shell elements
As an example of non-linear analysis, Fig. 4.37 shows a square plate simply supported on sides AB and CD. A transverse point load of 10.00 kN is applied at the

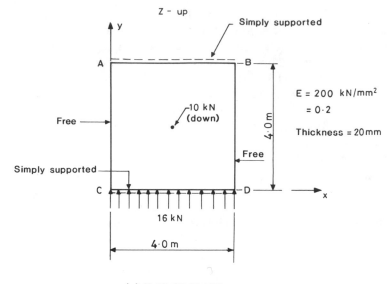

(a) PLAN OF PLATE

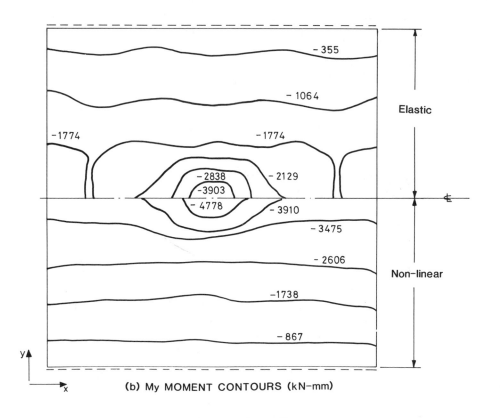

(b) My MOMENT CONTOURS (kN–mm)

Fig. 4.37 — Flat shell element example. (*Cont. next page.*)

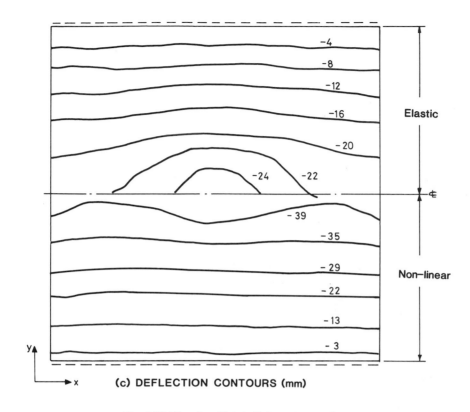

(c) DEFLECTION CONTOURS (mm)

Fig. 4.37 (*Cont.*) — Flat shell element example.

centre of the plate and a compressive in-plane load of 16 kN is applied on CD in the
y-direction.

Two analyses were carried out. Firstly elastic condtions were assumed and then
non-linear geometry effects were included in the model. Fig. 4.37(b) shows the M_y
moments predicted from the two runs. Note the increase in moment due to the non-
linear effect. The predicted lateral deformations are shown in Fig. 4.37(c). A
significant increase is also evident and the dishing effect at the centre in the elastic
case is flattened out in the non-linear run.

While such predictions may not be accurate representations of the real behaviour
of such a plate they do give an indication of the behaviour under non-linear
conditions. The design can be modified and the model re-run. The resulting
comparisons will be helpful in assessing the effectiveness of the changes.

5

Evaluation of results

In this chapter techniques for checking of computer runs are discussed.

Principle 2 in section 1.2 suggests that one should always be suspicious of the validity of computer results. In many cases it is not posssible to be sure that there are no errors and the time spent in checking the results only reduces the probability of error. However, one must take all reasonable steps to track down errors.

5.1 DATA CHECKING

The majority of errors are due to faults in the data which must be checked carefully. Screen and plotted graphics techniques are most helpful in this respect. Any attribute that can be shown graphically should represented in this way for checking.

Formal methods for checking data are not easily established. An item-by-item check is an obvious approach. If practicable this should be done by the person who set up the data and by someone else who was not previously involved.

Some data items can be checked directly from the results. The following actions are good practice:

- Check the sum of the reactions against the applied load. If they are not equal then it is likely that there is a data error, i.e. that the required loads have not been correctly specified. If the loads have been correctly specified, discrepancies of this type could be caused by ill-conditioning as discussed in section 5.2.
- Fixed values of deformation (e.g. restraints) should be checked in the results to ensure that the required values have indeed been imposed. Data errors are common in this connection especially where restraints are imposed at axes of symmetry.
- If there are symmetry conditions present then the corresponding symmetrical values should be be checked for equality.
- If constraints have been included then checks can be made to ensure that the geometric conditions have been imposed. Equality of deformations can easily be assessed but in some cases rigid body calculations involving rotations may be worthwhile.

• In frames, spot checks on joint equilibrium can be useful. This would appear to be only worthwhile for identifying ill-conditioning problems but I have found that such checks have led to identification of data errors.

5.2 SOLUTION CHECKS

The solution of a set of simultaneous equations always produces an approximation to the correct results. The basic source of such error is in 'truncation' where for example in multiplication half of the significant figures which can be generated have to be discarded. When this causes the results to be noticeably in error the equations are said to be 'ill-conditioned'.

Problems of ill conditioning arise from two sources:

(1) The structural system is poorly conditioned.
(2) Insufficient significant figures are used in the solution routine.

5.2.1 Ill-conditioned structures

A main cause of ill conditioning is large differences in the stiffnesses of parts of the structure. For example, a stiff structure on flexible supports can give problems in this direction. An example of this situation is given in MacLeod (1988). Note that the opposite situation — a flexible structure on stiff supports — tends not to suffer from such difficulties.

A cantilever with high span-to-depth ratio also tends to be ill conditioned. This is because small movements at the support can generate much greater movements towards the tip. In other words, the system is highly coupled. Less highly coupled systems, such as a vierendeel frame deforming exlusively in a shear mode (section 6.4), tend to give more stable equations.

Axial deformation in frames where bending is the dominant mode of member deformation can degrade the numerical stability. For example, axial deformation of the members of a vierendeel frame (section 6.4) normally does not affect behaviour significantly but can cause the system to be ill conditioned. In such situations it may be best to neglect the axial flexibility although this may not be easily achieved.

A useful strategy for helping to maintain a stable solution is to number the nodes from the most flexible part to the most rigid part (e.g. in a cantilever, start numbering at the tip and work towards the support). This rule would be used if a bandwidth solution is used (section 8.6.4). With a frontal solution one should number the elements from the flexible part to the stiff part.

5.2.2 Wordlength

The 'wordlength' is the number of binary digits used to store a floating point number. These digits are used to store the exponent and its sign plus the mantissa and its sign.

Typical single precision arithmetic uses 32 bit words giving 6 to 7 decimal significant digits for each number stored. This degree of precision can lead to ill conditioning problems, so using words based in 64 bits giving 11 to 12 significant decimal digits is recommended. This is the normal situation for double precision arithmetic but the term 'double precision' does not always mean 64 bit words.

In summary, with 32 bit words look out for ill conditioning; with 64 bit words ill conditioning is relatively uncommon.

5.2.3   Identification of ill-conditioning

Residuals of the solution

The simplest check for ill conditioning is to look at the residuals of the solution. The residuals are defined as follows.

The structural stiffness equations are written in the form:

$$\mathbf{P} = \mathbf{K}\Delta$$

where $\mathbf{P}$ is the vector of applied loads, $\mathbf{K}$ is the structural stiffness matrix and Δ is the vector of deformations, i.e. the solution vector.

The residual vector $\mathbf{R}$ is then:

$$\mathbf{R} = \mathbf{P} - \mathbf{K}\Delta \qquad\qquad\qquad (5.1)$$

Clearly all terms of R should be close to zero for a satisfactory solution. If not then ill conditioning is probably present. The converse is however *not* true. Low values in the residual vector do not guarantee that the solution is satisfactory. This is unfortunate since the residual vector is often relatively easy to calculate. For example, checking that the sum of the reactions is equal to the applied load or equilibrium spot checks are, in effect, checks on the residuals.

Decay of diagonal coefficients

Structural stiffness equations are normally solved by a direct method, the most popular being Gaussian elimination. This process involves 'reducing' the original matrix to a triangular form. A leading diagonal element of the reduced matrix is denoted by D_{ii}. The corresponding leading diagonal element in the original matrix is denoted K_{ii}. The ratio K_{ii}/D provides an *indication* of the number of significant figures that have been lost in the reduction process.

This ratio has the advantage that it does not require significant computer time to calculate it. It is not however a measure of accuracy. It only gives a warning signal.

Other Methods

There are techniques which can identify ill conditioning to a high degree of reliability but they require a significant amount of computer processing. They are therefore relatively expensive and seldom used in structural modelling.

5.3   CHECKING MODELS

When one is presented with a set of computer results and feels satisfied that the data have been adequately checked one then must decide whether or not they can be accepted. It is easy to pass them on directly to the next part of the design process but this should never be done without reflecting on their validity.

This can be done by looking at the results in relation to expectation. This expectation is described here as a 'checking model'. Such a model is based on the analyst's understanding of the behaviour of the system under consideration. It does not necessarily require calculations to be made.

5.3.1   Checks on the general form of the results

When one carries out an analysis one normally has some knowledge of the form of the results. One knows the likely direction of movement and has some idea of the level of stress. It is essential that such knowledge, imperfect as it may be, is put to use.

For example, for the frame with top point load shown in Fig. 5.1 one could look for the following features in the results:

- The overall lateral deformation would be close to a straight line (sec section 6.4).
- The lateral deformation at A would be slightly greater than at B due to beam axial deformation
- The results would be roughly skew-symmetrical about axis *a–a*.
- Towards the middle of the frame the end moments in the columns would tend to be equal at the top and the bottom (giving mid height points of contraflexure).

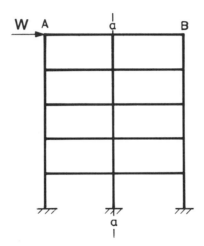

Fig. 5.1 — Frame with lateral point load.

We do have good knowledge about the behaviour of such frames but normally less insight than this will be available. However, there are always some features that are susceptible to this type of treatment.

Another technique is to look for discontinuities in results. If they do not correspond to discontinuities in the loading or in the structure then they may point towards an error.

5.3.2   Checking calculations

A final step in checking is to do some calculations. The models used for this would normally be simpler than that being checked but must reflect the required behaviour to an adequate degree. 'Back of an envelope' calculations can be valuable although simplified computer models may also be desirable especially for checking big important runs.

Some checking models are described in Chapter 6.

5.3.3 Relationship between results from different models

If the results from a checking model do not agree with those from that being checked then at least one of the models has significant errors and further investigations are needed. One has now to decide which model is more likely to be in error.

A distinct possibility is that the checking model is not sufficiently accurate or that it is based on unsound knowledge. For example, some years ago we were asked to investigate the effect of poor support conditions on the stresses in a facade wall of the type shown in Fig. 5.2(a). We used plane stress elements and removed a length of support as shown. The results gave horizontal compression at point A and our initial checking model was that the wall should act as a cantilever. This would give horizontal tension at A. The wall does cantilever but our error was to assume that this was the dominant mode of behaviour. The dominant mode is a shearing type of deformation as shown in Fig. 5.2(b). The horizontal compression at A was thus explained and, in the process, our knowledge of the behaviour of such systems improved significantly.

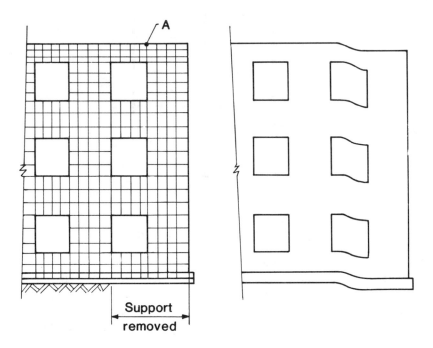

(a) PLANE STRESS ELEMENT MODEL   (b) SHEARING DEFORMATION

Fig. 5.2 — Settlement of a facade wall.

When comparing the results from two models one hopes to get a good correlation. Results that are very close should be treated with suspicion since both results could be wrong but happen to give similar answers. The probability of this occurring ought to be low but I have experienced a number of situations of this type.

Also a correlation for a particular parameter may be valid but other parameters may not be correspondingly accurate. In particular, deformation estimates tend to be much better than those for stress or member forces as discussed in section 1.6.

5.4 SOFTWARE AND HARDWARE ERRORS

Like accuracy of data, one's confidence in a computer program to carry out tasks without error is a matter of probability. It is not possible to ensure that a program is completely error free. Therefore the possiblity of errors due to program faults cannot be ruled out. Most standard analysis packages have been checked extensively by use and one would only suspect software faults as a second last resort.

The last resort is to suspect a hardware fault. Normally these are catastrophic and entirely evident. However, hardware faults can cause errors which are not easily identified. This can, for example, result from overheating of the computer causing problems at the insignificant end of arithmetic manipulations leading to accumulated error in the results. Situations of this nature can be very difficult to identify but are fortunately rare.

6

Checking models

This chapter describes some models that can be solved with a minumum of calculation. They are presented here as checking models for assessing the validity of computer results as described in section 5.3.2. They can also be used in some circumstances directly in the design process and, in accordance with principle 1 in section 1.2, they should be used in this way if this is practicable. Circumstances which would allow their direct use include:

- Where the general level of stress and deformation is well below the limit state levels.
- Where the simpler model is known to give adeqate accuracy in the given situation.

6.1 BENDING ACTION

Bending theory is one of the most widely used models in structural mechanics and can be most useful for checking purposes. The use of bending to model from behaviour is discussed in sections 6.2.2, 6.4 and 6.7.2.

Where the structure (or part of the structure) has a relatively low length-to-depth ratio shear deformation needs to be taken into account as illustrated in Case study 6.1.

Table A8 gives some useful formulae for bending and shear in beams.

Case study 6.1   Cantilever bracket
Fig. 6.1(a) shows a steel cantilever bracket fixed to a column. Fig. 6.1(b) shows a finite element model. The web is treated as plane stress elements and the flanges as beam elements using FLASH (Walder & Green 1981). The results give deflections under the load points of 0.5301 and 0.6365 mm. The axial stress in the beam element at A is 76.2 N/mm^2. To check these values, the system is treated as a cantilever with tip load 200 kN, length 530 mm.

(1) Estimate deflection at centre of load

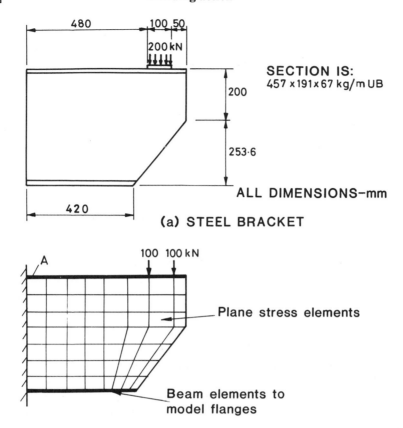

ALL DIMENSIONS–mm

(a) STEEL BRACKET

(b) PLANE STRESS MODEL

Fig. 6.1 — Cantilever bracket.

Bending deflection $= \dfrac{WL^3}{3EI} = \dfrac{200 \times 530^3}{3 \times 200 \times 2.93 \times 10^8} = 0.169 \text{ mm}$

Shear deflection $= \dfrac{WL}{AG} = \dfrac{200 \times 530}{8.5 \times 453.6 \times 77} = 0.357 \text{ mm}$

$$\text{Total} = 0.526 \text{ mm}$$

Average value for load points from finite element = 0.583 mm

Note that shear deformation dominates the deflection.

(2) Estimated bending stress at A

$$\sigma_A = \frac{WL}{Z} = \frac{200 \times 530}{1293 \times 10^3} = 81.9 \text{ N/mm}^2$$

Plane stress element model value = 76.2 N/mm^2.

The comparison is therefore favourable.

6.2   TRUSS ACTION

The simplest assumption that one can make about structural behaviour is that the stress or the strain is constant and unidirectional over a region. This is commonly done in reinforced concrete for shear reinforcement and examples of such a model are discussed in section 4.2.6 for the bracing action in infilled frames and for the stiffening effect of cladding of buildings in section 4.7.4.

Early approaches to finite element modelling of plane stress situations used equivalent truss systems (Hrennikoff 1941, McHenry 1943). The converse concept of treating a truss system as a continuous medium can also be useful. The equivalent beam described in section 6.2.2 is an example of this.

6.2.1   Forces in truss members

Trusses are often statically indeterminate or can be modelled as such by removing compression diagonals in cross braced systems. Selected member forces can then be estimated on the basis of equilibrium. Ignoring the compression diagonals will cause an overestimate of force in the other truss members and in deflection.

6.2.2   Parallel chord trusses

A parallel chord truss has a structural action analogous to that of a beam. The top and bottom chords are equivalent to the flanges while the posts and diagonals are equivalent to the web. This observation leads to the concept of treating the truss as an equivalent beam. Figure 6.2(a) shows a parallel chord bridge truss and Fig. 6.2(b) is a beam equivalent. The properties of the beam are:

(1) Equivalent bending stiffness

$$EI_e = EA_c b^2/2 \qquad\qquad (6.1)$$

where A_c is the area of the chord members assumed to be equal top and bottom. If they are not equal, I_e can be calculated as the second moment of the chord areas about the centroid of the chord areas. That is, a 'neutral axis' within the depth of the truss needs to be established about which the second moments of area are based.

> b is the depth of the truss
> E is the E value of the truss material

(2) Equivalent shear stiffness

$$(\bar{A}G)_e \text{ (equivalent shear stiffness)}$$
$$= EA_d f_1 \cos\theta \sin^2\theta \qquad\qquad (6.2)$$

where A_d is the area of a diagonal member, θ is the angle of a diagonal member to the horizontal, $f_1 = 1.0$ for single bracing $= 2.0$ for cross bracing (where the diagonals can sustain compressive load) $= 0.5$ for a K-braced truss.

Use of equations (6.1) and (6.2) together with the deflection formulae of Table A8 normally give estimates of truss deflection to a useful degree of accuracy.

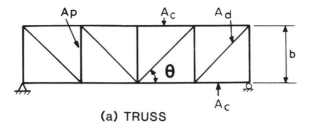

(a) TRUSS

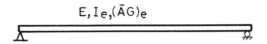

$E, I_e, (\bar{A}G)_e$

(b) BEAM EQUIVALENT

Fig. 6.2 — Beam element of a parallel chord truss.

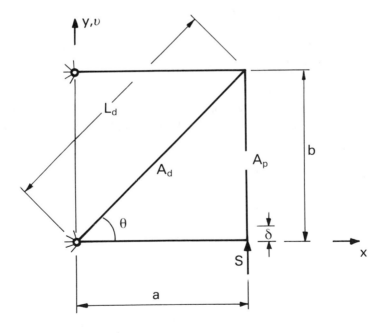

Fig. 6.3 — Truss bent.

Derivation of equation (6.2)

Equation (6.2) is derived as follows: consider the bent of a truss shown in Fig. 6.3.
The deflection δ in the line of shear S is

$$\delta = \frac{SL_d}{EA_d\sin^2\theta} + \frac{Sb}{EA_p} \tag{6.3}$$

where A_d and A_p are the areas of the diagonal and post members respectively, a, b, L_d are the dimensions as shown in Fig. 6.3, $\frac{\delta}{a}$ is the slope of the lateral displacement of a truss.

Treating the truss as being in the xy plane, as shown in Fig. 6.3, this slope can be defined as

$$\frac{\delta}{a} = \frac{d\upsilon}{dx}$$

Substituting this into equation (6.3) and rearranging gives:

$$S = \frac{a}{\left[\dfrac{L_d}{EA_d\sin^2\theta} + \dfrac{b}{EA_p}\right]} \frac{d\upsilon}{dx} \tag{6.4}$$

In comparison with equation (2.10) the factor in square brackets before $d\upsilon/dx$ in equation (6.4) can be defined as an equivalent shear stiffness. For checking calculations the post flexibility can be ignored and the bracketed factor reduces to the equivalent shear stiffness quoted in equation (6.2).

Bracing trusses in buildings

The same technique can be used for bracing trusses in buildings. For a multibay truss of this type — Fig. 6.4 — the equivalent bending stiffness is based on all the column

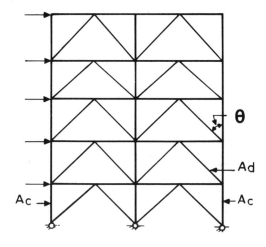

Fig. 6.4 — K-braced truss.

areas and the equivalent shear stiffness is the sum of the values given by equation (6.2) for each braced bay. In a K-braced truss the diagonals contribute to the bending mode behaviour. This effect can be modelled by adding an area $A_d \cos^3 \theta$ to the column areas.

Equivalent beam element

A parallel chord lattice truss in a structural system can be treated as an equivalent beam element using equations (6.1) and (6.2). The cross-section area of the equivalent beam can be taken as the sum of the areas of the chord members.

6.2.3   Truss models of membrane systems

The opposite strategy to using an equivalent shear stiffness in a beam model of a truss is to treat shear panels as part of a truss model as discussed in section 4.7.4. In this section a justification for equation (4.10) is presented.

Equation (4.9) is a flexibility relationship for a shear panel (Fig. 4.35). Transforming equation (4.9) to a stiffness relationship of the form of equation (2.10) gives:

$$S = \frac{a}{c}\frac{dv}{dx} \tag{6.5}$$

The equivalent beam shear stiffness for a shear panel is

$$(\bar{A}G)_e = \frac{a}{c} \tag{6.6}$$

The shear stiffness for a diagonally braced panel based on equation (6.2) with $f_1 = 1.0$ and $\cos\theta = a\sin\theta/b$ is:

$$(\bar{A}G)_e = \frac{EA_d a \sin^3 \theta}{b} \tag{6.7}$$

Equating (6.6) and (6.7) gives:

$$A_d = \frac{b}{E\sin^3\theta c} \tag{6.8}$$

A_d is the equivalent area of the tension diagonal to represent the shear stiffness of a panel.

Equation (6.8) is the same as equation (4.10). Davies & Bryan (1982) quote this expression for use with infill panels to brace multistorey structures.

6.3   ARCH ACTION

Fig. 6.5(a) shows the force actions in an arch. The main arch force is assumed to be compressive, giving rise to horizontal and vertical reaction components at each end. The function of an arch is the same as a beam in that it concentrates load to its supports, i.e. it redistributes vertical load laterally. A main difference is that it requires a horizontal thrust at the supports. This is either compressive from outside the arch or from a tie across the bottom of the arch.

Another arch situation is in shear walls which are supported on columns (section

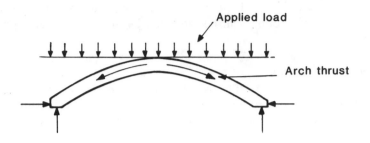

(a) ARCH FORCES

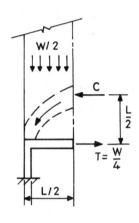

(b) COLUMN SUPPORTED
   SHEAR WALL

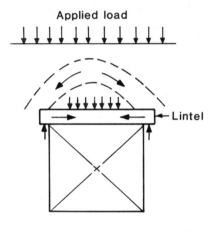

(c) LINTEL ACTION

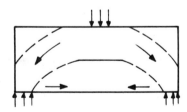

(d) ARCH ACTION AT BEAM SUPPORTS

Fig. 6.5 — Arch action.

4.3.2). In such a case the arch forces can be roughly estimated by considering equilibrium of a free body diagram of half of the arch taking the height to the centre-line of the arch as half the span as shown in Fig. 6.5(b) (Green 1972).

Arch action occurs in lintel beams above openings in walls. The lintel acts more as a tie to the arch than as a bending element — Fig. 6.5(c).

The ends of a simply supported beam has arch type action where the compression zone slants downward to the reaction area — Fig. 6.5(d)

Dome action is a three-dimensional form of arch action. Fig. 6.6 is a vertical section through a dome. The thrust from the compressive forces in the shell of the dome is transferred into a lower ring beam in tension and an upper compressive ring beam (if present). The hoop forces will therefore vary from being tensile at the top through to compressive at the base. A pyramid shape will exhibit the same type of behaviour.

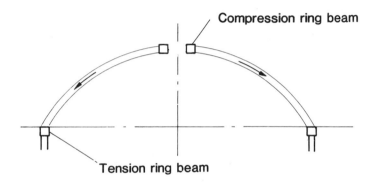

Fig. 6.6 — Section through a dome.

6.4 VIERENDEEL FRAMES

A vierendeel frame has rectangular panels with no cross bracing. It is not an efficient means of transmitting transverse load but it is sometimes used to resist lateral load in buildings and for architectural reasons when cross bracing is undesirable.

Fig. 6.7(a) shows a typical unbraced rigid jointed building frame of this type. For a simplified lateral load analysis of such a frame the first step is to reduce the frame to a single bay equivalent as described in section 4.7.2 — see Fig. 6.7(b).

The portal method assumes that there are points of contraflexure at mid-height of columns and mid-span of beams of such a frame. The applied shear is distributed equally to the columns and hence a statically determinate equivalent frame is produced as shown in Fig. 6.7(c). Figure 6.7(d) shows a bent from this frame which illustrates how the internal actions can be calculated. S is the total shear applied to the frame above the beam level being considered.

The mid-span assumption for points of contraflexure in the beam is not accurate unless the frame is 'proportioned'. A proportioned frame can be divided into a set of

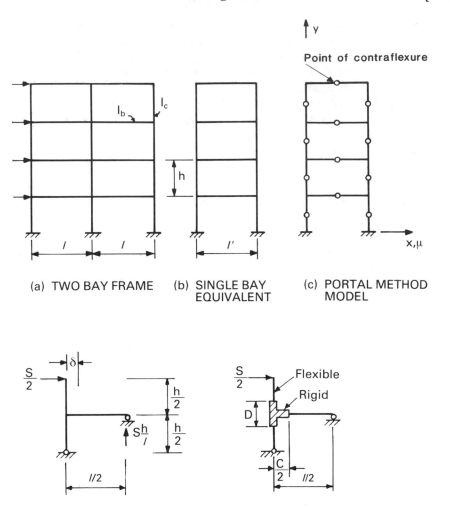

(a) TWO BAY FRAME   (b) SINGLE BAY   (c) PORTAL METHOD
                         EQUIVALENT   MODEL

(d) SINGLE STOREY BENT   (e) BENT WITH RIGID
                              CONNECTION

Fig. 6.7 — Vierendeel frame models.

equivalent single bay frames each of which has the same column to beam stiffness ratio (λ). Real frames probably seldom conform to this condition but this does not invalidate the use of the portal method for checking.

The validity of the mid height position for the column points of contraflexure depends on the value of the ratio

$$\lambda = \frac{I_c/h}{I_b/l}$$

where I_c, I_b are the column and beam I values respectively of the single bay equivalent frame, h is the storey height, l is the span of the beam of the single bay frame.

When this ratio is small, i.e. significantly less than 1.0, then the joint rotation is low and the column points of contraflexure will be close to mid-height. When λ is high, i.e. greater and 5.0, the degree to which the beams transfer vertical shear becomes less significant and the column points of contraflexure will not be close to mid-height. Thus below $\lambda = 5$, the assumption tends to give an order of magnitude estimate of column moment improving as λ decreases. The error is most noticeable at the base of the columns where, unfortunately, it is most important.

The deflection of the portal method model shown in Fig. 6.7(c) can be calculated as follows:

The deflection δ of the bent of Fig. 6.7(d) is given by:

$$\delta = \frac{Sh^3}{12E\Sigma I_c}[1 + 2\lambda] \qquad (6.9)$$

where ΣI_c = sum of the I values for all columns of the frame.

δ/h is the slope of the lateral displacement of the frame. Treating the frame as being in the xy plane as shown in Fig. 6.7(c), this slope can be defined as:

$$\frac{\delta}{h} = \frac{du}{dx}$$

Substituting this into equation (6.9) and rearranging gives:

$$S = \frac{12E\Sigma I_c}{h^2[1 + 2\lambda]}\frac{du}{dx} \qquad (6.10)$$

Equation (6.10) has the same form as equation (2.10). Therefore the frame of Fig. 6.7(c) tends to deform laterally in a shear mode.

This is an important observation for checking of such frames. If the shear is constant, i.e. with top point load, du/dx is constant. Therefore the deflected shape is straight — Fig. 6.8. With uniformly distributed lateral load the deflected shape is parabolic — Fig. 6.8. Insight into this type of behaviour may be gained from Fig. 3.7(b). Its significance in relation to interaction with walls is discussed in section 6.6.2.

Comparing equations (2.10) and (6.10), an equivalent shear stiffness for the frame can be established as:

$$(\overline{A}G)_c = \frac{12E\Sigma I_c}{h^2[1 + 2\lambda]} \qquad (6.11)$$

Thus the deflection of a vierendeel frame can be estimated using the beam deflection formulae given in Table A8 using the equivalent shear stiffness of equation (6.7).

There is a bending mode component with such frames due to axial deformation of the longitudinal members. An equivalent bending stiffness can be used as in equation (6.1). (This effect tends to be significant only with tall building frames.) A vierendeel

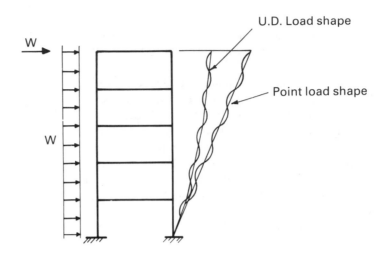

Fig. 6.8 — Lateral deformation of a vierendeel frame.

frame can therefore be treated as an equivalent beam element of the type described in section 6.2.2.

The finite sizes of the beams and of the columns can be considered in the shear stiffness. The relevant bent is shown in Fig. 6.7(b).

The equivalent of equation (6.9) for this bent is

$$\delta = \frac{Sh^3}{12E\Sigma I_c}[(1 - \beta_D)^3 + 2\lambda(1 - \beta_C)^3] \qquad (6.12)$$

where $\beta_D = D/h$, $\beta_C = C/l$, D = beam depth, C = column width.

The corresponding shear stiffness is:

$$(\bar{A}G)_e = \frac{12E\Sigma I_e}{h^2[(1 - \beta_D)^3 + 2\lambda(1 - \beta_C)^3]} \qquad (6.13)$$

In taller frames the column stiffness tends to vary with height. Equation (6.10) can be modified to take account of a linear decrease with height (MacLeod 1971) but for checking models, an average I value is probably adequate.

Equations (6.11) or (6.13) in a beam model can be useful for assessing the effect of structural modifications on frame stiffness.

6.5 SHEAR WALLS WITH OPENINGS

6.5.1 The continuous connection method

The line of openings of the frame model of Fig. 4.5(b) can be treated as a continuous connection between the 'columns' of the frame (Beck 1962, Rosman 1964). The problem can then be formulated as a second order differential equation and solved for different load cases and boundary conditions. For simple cases where formal solutions for the differential equation are possible this 'continuous connection' model provides a useful way of solving the frame model. It is best to write a program to do the calculations but for a quick check for a wall with a single row of openings and fixed base the following approach can be used to estimate lateral deflection.

Estimation of lateral deflection
The top lateral deflection of a wall with a single row of openings (Fig, 6.9) can be

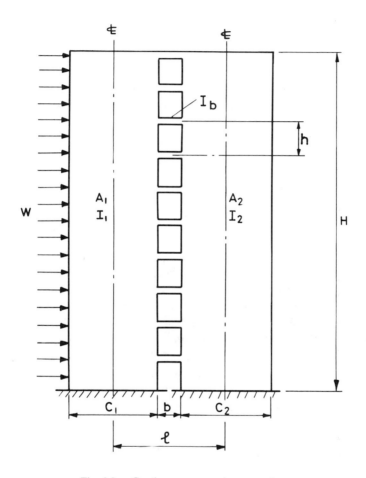

Fig. 6.9 — Continuous connection example.

calculated using

$$\Delta_{top} = R \frac{WH^3}{EI'} F \qquad\qquad (6.14)$$

where

R is a load coefficient having values:

Top lateral point load $R = 1/3$
Uniformly distributed lateral load $R = 1/8$
Triangular (earthquake) load $R = 11/60$

W is the total load, H is the total height, I' is the second moment of area of the wall cross-section, F is the flexibility factor from Table 6.1.

The independent variables in Table 6.1 are

$$\alpha H = H \sqrt{\frac{12 I_b l^2}{h b^3} \left[\frac{1}{\Sigma I} + \frac{1}{B} \right]} \tag{6.15}$$

$$\mu = 1 + \Sigma I / B$$

where b, h and l are defined in Fig. 6.9, I_b is the second moment of area of a connecting beam $\Sigma I = I_1 + I_2$, I_1 and I_2 are the second moments of area of columns 1 and 2,

$$B = \frac{l^2}{\left[\dfrac{1}{A_1} + \dfrac{1}{A_2} \right]}$$

A_1 and A_2 are the areas of columns 1 and 2.

$$I' = \Sigma I + B$$

The flexibility factor F is the amount by which the openings increase the flexibility of the wall. With $F = 1.0$ there is full shear transfer across the line of openings, i.e the openings do not reduce the stiffness of the wall.

The relationship between αH and wall stiffness is readily appraised using Table 6.1. With high values of αH ($aH > 8$) the wall stiffness is not significantly reduced by the openings. With low αH ($\alpha H < 4$) the openings radically degrade the stiffness.

Case study 6.2 — Wall with single row of openings
The wall shape used in Fig. 6.9(a) is used to illustrate the use of equation (6.14). The top lateral deflection Δ_{top} is to be estimated. The dimensions are:

Storey height $h = 3.0\,\text{m}$
Number of stories $= 10$
Total height $H = 30.0\,\text{m}$
Wall widths $C_1, C_2 = 8.0, 8.0\,\text{m}$
Opening width $b = 2.0\,\text{m}$
Wall thickness $t = 200\,\text{mm}$
Lintel beam depth $= 500\,\text{mm}$
Uniformly distributed
    lateral load $= 6.0\,\text{kN/m}$

$I_1 = I_2 = 0.2 \times 8^3/12 = 8.53\,\text{m}^4$
$A_1 = A_2 = 0.2 \times 8 = 1.6\,\text{m}^2$
$I_b = 0.2 \times 0.5^3/12 = 20.8 \times 10^{-4}\,\text{m}^4$
$l = 10, \quad B = 1.6^2 \times 10^2/3.2 = 80$

Table 6.1 — Values of flexibility factor F for Equation (6.14)

Loading	μ	αH														
		0	0.5	1	2	3	4	5	6	7	8	9	10	12	15	20
Point	1.1	11.0	10.1	8.15	4.88	3.22	2.41	1.96	1.69	1.52	1.41	1.33	1.27	1.19	1.12	1.07
	1.2	6.0	5.55	4.58	2.94	2.11	1.70	1.48	1.34	1.26	1.20	1.16	1.13	1.09	1.06	1.03
	1.3	4.33	4.03	3.38	2.29	1.74	1.46	1.32	1.23	1.17	1.13	1.10	1.09	1.06	1.04	1.02
	1.4	3.50	3.27	2.78	1.97	1.55	1.35	1.24	1.17	1.13	1.10	1.08	1.06	1.047	1.031	1.018
U.D.	1.1	11.0	10.1	8.2	5.03	3.38	2.55	2.09	1.80	1.62	1.49	1.40	1.33	1.23	1.16	1.09
	1.2	6.0	5.55	4.61	3.02	2.19	1.76	1.54	1.40	1.31	1.24	1.20	1.16	1.12	1.08	1.04
	1.3	4.33	4.03	3.41	2.34	1.80	1.52	1.36	1.27	1.21	1.16	1.13	1.11	1.08	1.05	1.03
	1.4	3.50	3.28	2.80	2.01	1.60	1.39	1.27	1.20	1.15	1.12	1.10	1.08	1.06	1.04	1.02
Triangular	1.1	11.0	10.1	8.2	5.0	3.34	2.51	2.05	1.77	1.59	1.46	1.38	1.31	1.22	1.15	1.08
	1.2	6.0	5.55	4.6	3.0	2.17	1.76	1.53	1.38	1.29	1.23	1.19	1.16	1.11	1.07	1.04
	1.3	4.33	4.03	3.40	2.33	1.78	1.50	1.35	1.26	1.20	1.15	1.13	1.10	1.07	1.05	1.03
	1.4	3.50	3.28	2.80	2.00	1.59	1.38	1.26	1.19	1.15	1.12	1.09	1.08	1.055	1.036	1.021

$$I' = 2 \times 8.53 + 80 = 97.1$$
$$\mu = 1 + 97.1/80 = 1.213$$

$$\alpha H = 30 \sqrt{\frac{12 \times 20.8 \times 10^{-4}}{3 \times 2^3} \left[\frac{1}{17.06} + \frac{1}{80}\right]} = 2.58$$

$$F = 2.48 \quad R = \tfrac{1}{8}$$
$$W = 30 \times 6 = 18 \text{ kN}$$

$$\Delta_{\text{top}} = \frac{1}{8} \times \frac{180 \times 30^3}{20 \times 10^6 \times 97.1} 2.48 \times 10^3$$

$$= 7.76 \text{ mm}$$

Note that the low value of αH and the correspondingly high value of F (in comparison with 1.0) show that the openings significantly reduce the stiffness of the wall.

Internal actions
With high αH (or low F) the wall can be treated as a single cantilever and the stresses estimated assuming full transfer of moment across the wall, e.g. for the wall of Fig. 6.9 the vertical stress at the base due to the lateral load W would be approximated by

$$\sigma = \frac{M}{Z} = \frac{WH/2}{[I'/(C_1 + b/2)]} \tag{6.16}$$

With low αH, as in the example in this section, the internal actions need to be calculated using expressions derived from solutions for the continuous connection model, e.g. from Irwin (1984). Useful curves for this purpose are also given in Coull & Choudhury (1967a, b) and Irwin (1984).

Equivalent wall
One can treat a wall with a row of openings as an equivalent plain cantilever using the flexibility factors F from Table 6.1. The equivalent I value I_e is based on

$$I_e = \frac{I'}{F} \tag{6.17}$$

The shear area of the equivalent cantilever is the sum of the shear areas of the two columns of the model.

Solutions for other wall configurations
The continuous connection model was devised at a time when computer processing was much less available than now. Many papers give solutions for different load cases and boundary conditions. Except for the simple case illustrated in Fig. 6.9, the use of a computer to do the calculations is desirable (though not essential in some cases). A general approach to formulating such problems is given in Balendra et al. (1984).

6.6   DISTRIBUTION OF LATERAL LOAD IN BUILDINGS

6.6.1   Rigid plate on springs model

The rigid plate on springs model discussed in section 4.7.3 can be treated as a
checking model.

No torsion

If torsion is neglected the system can be treated as a rigid beam on spring supports —
Fig. 6.10. The distribution of load to the supports is then in direct proportion to their
top stiffnesses k_i.

In this case the stiffness relationship for the system is:

$$W = \Sigma k_i \Delta \tag{6.18}$$

where Δ is the top deflection which can be calculated using:

$$\Delta = \frac{W}{\Sigma k_i} \tag{6.19}$$

The load on a support frame is

$$P_i = k_i \Delta = \frac{K_i}{\Sigma k_i} W \tag{6.20}$$

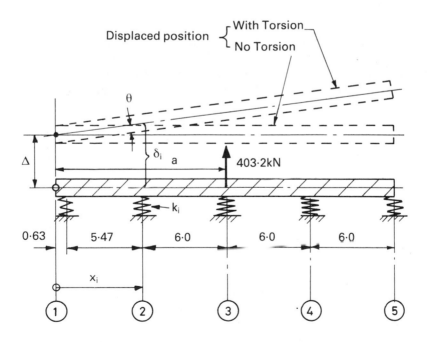

Fig. 6.10 — Rigid plate on springs example.

With torsion (two degrees of freedom)

If torsion is considered then an extra freedom (corresponding to a rotation) needs to be added to the rigid beam on springs — Fig. 6.10. The system deformations are Δ and θ. The corresponding deformation of a spring at distance x_i from the origin is

$$\delta_i = [l \ x_i] \begin{Bmatrix} \Delta \\ \theta \end{Bmatrix} \tag{6.21}$$

The equilibrium equations are then

$$\begin{Bmatrix} W \\ M \end{Bmatrix} = \Sigma \begin{bmatrix} l \\ x_i \end{bmatrix} P_i \tag{6.22}$$

where M is the moment of the applied loads about the origin. ($M = Wa$ for the system of Fig. 6.10.)

The load-deformation relationship of a spring is

$$P_i = k_i \delta_i$$

i.e. $\quad P_i = k_i(\Delta + x_i\theta) \tag{6.23}$

Substituting equation (6.23) into equation (6.22) gives

$$\begin{Bmatrix} W \\ M \end{Bmatrix} \begin{bmatrix} \Sigma k_i & \Sigma k_i x_i \\ \Sigma k_i x_i & \Sigma k_i x_i^2 \end{bmatrix} \begin{Bmatrix} \Delta \\ \theta \end{Bmatrix} \tag{6.24}$$

Equation (6.24) can be solved to get the system deformations Δ and θ from which the spring forces are obtained using equation (6.23). The solution in this case can be conveniently programmed using a spreadsheet.

For hand solution it is helpful to use the 'shear centre' of the spring system as the origin. If this is done then the off-diagonal terms of the matrix of equation (6.24) are zero and the solution of the equations is simplified. The shear centre is found by taking first moments of spring stiffness about any point to find the position of the resultant of spring stiffness which is the shear centre.

Treatment of the plate on springs model with three degrees of freedom is discussed in section 8.9.

Case study 6.3   Four-storey building

The four-storey building under lateral load considered in Case study 4.6 (page 108) is re-analysed using the procedures outlined in section 6.6.1.

No torsion

The rigid beam on springs model with no torsion (Fig. 6.10) is denoted here as 'M5'. This is a simplified form of Model M1 of Table 4.5 (Fig. 4.31(a)). The stiffnesses for the springs are as quoted in Case study 4.6.

For M5 the top deflection is obtained using equation (6.19), i.e.

$$\Delta = \frac{W}{\Sigma k_i} = \frac{403.2}{(431.6 + 3 \times 9.93 + 113.5)} = \frac{403.2}{574.9} = 0.7013$$

The load in frame 1, for example, is given by

$$p_1 = \frac{k_1}{\Sigma k_i} W = \frac{431.6}{574.9} \times 403.2 = 302.7 \text{ kN}$$

and

$$p_5 = \frac{k_5}{\Sigma k_i} W = \frac{113.5}{574.9} \times 403.2 = 79.6 \text{ kN}$$

To calculate the moment in the core of frame 1 the results of the computer analysis of that frame on its own (see typical frame of this type in Fig. 4.32(c)) is multiplied by the factor:

$$k_1/\Sigma k_i = 431.6/574.9 = 0.751$$

The moment at the base of the core in the analysis for frame 1 was 2704 kN m. Therefore the estimate of actual base moment is $2704 \times 0.751 = 2031$ kN m.

Table 6.2 gives a comparison between the results for M1 and M5. The comparison

Table 6.2 — Four-storey building-lateral load with no torsion

Model	Figure number	Top displacement in y-direction (mm)		Base moment (kN m)	
		Core of frame 1	Central wall of frame 5	Core of frame 1	Central wall of frame 5
M1	4.31(a)	0.676	0.671	2082	428
M5	6.10 (no torsion)	0.689	0.698	2031	479

is favourable mainly because the walls in frames 1 and 5 dominate the behaviour and therefore the effect of shear wall-frame interaction is low. M5 neglects the fact that the walls and the frames have basically different modes of lateral displacement.

With torsion
Model M4 of Table 4.5 (Fig. 4.33(c)) and that of Fig. 6.10 with torsion are the same.

The hand solution can be laid out in tabular form as in Table 6.3. A spreadsheet system is useful for doing this.

Substituting from Table 6.3 into equation (6.24) gives

$$\begin{Bmatrix} W \\ M \end{Bmatrix} = \begin{Bmatrix} 401.6 \\ 401.6 \times 12 \end{Bmatrix} = \begin{bmatrix} 574890 & 3353388 \\ 3353388 & 70552022 \end{bmatrix} \begin{Bmatrix} \Delta \\ \theta \end{Bmatrix}$$

solving this gives:

$$\Delta = 0.4169 \text{ mm} \quad \theta = 0.0000488 \text{ radians}$$

Back-substituting into equation (6.21) and (6.23) gives the P_i and δ_i columns respectively. The results for δ_i are the same as quoted for M4 in Table 4.5.

Table 6.3 — Beam on springs with torsion

Spring	k_i kN/m	x_i (m)	$k_i x_i$	$k_i x_i^2$	P_i (kN)	δ_i (mm)
1	431600	0.63	271908	171302	193.2	0.4476
2	9930	6.0	59580	357480	7.0	0.7095
3	9930	12.0	119160	1429920	9.9	1.0021
4	9930	18.0	178740	3217320	12.9	1.2946
5	113500	24.0	2724000	65376000	180.2	1.5878
	574890		3353388	70552022	403.2	

6.6.2 Shear wall frame interaction

The single equivalent wall and single equivalent frame concept developed in section 4.7.2 is used to develop a checking model. When lateral load is applied to such a system — Fig. 6.11(a) — the wall tends to deform dominantly in a bending mode (provided the height-to-width ratio is not low). The frame tends to deform in a shear mode, and when the two are forced to take up the same lateral deflection at each storey level, non-uniform interacting forces occur.

In the direction of loading the wall pushes on the frame at the top and the frame pushes on the wall at the base. Thus one can find that the shear at the base of the wall is greater than the applied shear and, correspondingly, the shear at the base of the frame is in the opposite direction to the applied shear. If one was not aware of this behaviour one might readily take such a pattern of shears to be in error.

Calculation model
Fig. 6.11(b) shows a typical distribution of shear on the frame of the model of Fig. 6.11(a). As a first approximation this is assumed to be constant. Thus the frame takes a top point load. The model of Fig. 6.11(a) therefore reduces to that of Fig. 6.11(c) where the interaction between the frame and the wall is only at the top of the system (MacLeod 1971). This model can be further simplified to that of a wall with a top spring support — Fig. 6.11(d). The system of Fig. 6.11(d) can be readily solved since it is singly statically indeterminate. With the interacting load at the top of the frame denoted as P and the total load on the system as W, then Table 6.4 gives values for the ratio P/W for three different load cases. In Table 6.4 the frame stiffness K_f is defined as the top lateral point load to cause unit deformation in its line of action. The wall stiffness K_w is also the top point load stiffness.

Accuracy of this model is discussed in MacLeod (1971) and Green (1987). The maximum shear on the frame tends to be underestimated by the order of 30% and results are poor when the frame is stiffer than the wall. Otherwise the model predicts the interaction effect to a degree of accuracy which would normally be acceptable for checking calculations.

Case study 6.4
Model M2 of Fig. 4.31(b) and Table 4.5 is solved using the relevant relationship from Table 6.4. Relevant member properties are quoted in Case study 4.6 (page 108).

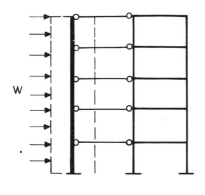

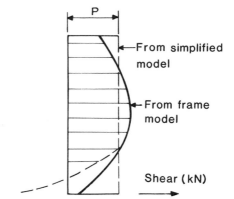

(a)SHEAR WALL FRAME MODEL   **(b) SHEAR FORCE ON FRAME**

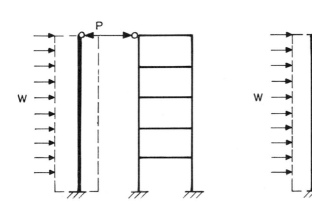

(c) INTERACTION AT TOP ONLY   **(d)WALL WITH SPRING SUPPORT
                                          MODEL**

Fig. 6.11 — Shear wall frame interaction.

The calculations are:

(1) Establish K_w. This is the top stiffness relative to a top lateral *point* load.

$$\text{i.e.} \quad K_w = \frac{3EI_w}{H^3} = \frac{3 \times 20 \times 10^6 \times 8.748}{14^3}$$

$$= 0.1913 \times 10^6 \text{ kN/m}$$

Table 6.4 — Ratio P/W for shear wall frame interaction model

Loading type W = total load	Ratio $\dfrac{P}{W}$
Point load at top	$\dfrac{P}{W} = \dfrac{1}{1 + \dfrac{K_w}{K_f}}$
Uniformly distributed	$\dfrac{P}{W} = \dfrac{3/8}{1 + \dfrac{K_w}{K_f}}$
Triangular (earthquake)	$\dfrac{P}{W} = \dfrac{11/20}{1 + \dfrac{K_w}{K_f}}$

(Shear deformation can be included in K_w by defining it as

$$K_w = \sum \frac{1}{\left[\dfrac{H^3}{3EI_w} + \dfrac{H}{\overline{A}G}\right]}$$

where $\overline{A}G$ is the shear stiffness of a wall.)

(2) Establish K_f — also the top point load stiffness.

The frame deforms in a shear mode and therefore the top deflection (from Table A8) is

$$\Delta = \frac{WH}{(\overline{\overline{A}G})_e}$$

$$\therefore K_f = \frac{W}{\Delta} = \frac{(\overline{A}G)_e}{H}$$

From equation (6.11) $(\overline{A}G)_e = \dfrac{12E\Sigma I_c}{h^2[1 + 2\lambda]}$

Hence $K_f = \dfrac{12E\Sigma I_c}{h^2[1 + 2\lambda]H}$

$$\lambda = \frac{1}{2}\frac{I_c/h}{I_b/l} = \frac{1}{2}\frac{0.0189/3.5}{0.0573/5.0} = 0.236$$

$$\therefore K_f = \frac{12 \times 20 \times 10^6 \times 0.0189}{3.5^2 \times 1.472 \times 14} = 0.0180 \times 10^6 \text{ kN m}$$

(3) Calculate the Interaction Force P

For the U.D. load case, from Table 6.4

$$\frac{P}{W} = \frac{3/8}{1 + K_w/K_f}$$

$$\therefore P = \frac{0.375 \times 403.2}{1 + 0.1913/0.0180} = 13.0 \text{ kN}$$

The sums of the frame column shears at each storey working down from the top from the computer output for Model M2 are 18.1, 15.9, 13.2, 6.3 kN.

If P is multiplied by 1.3 (i.e. $P = 13.0 \times 1.3 = 16.9$ kN) to take account of the typical underestimate by 30% discussed in section 6.6.2 then the checking model of section 6.6.2 gives a reasonable estimate of maximum frame shear.

(4) Calculate top deflection

$$\Delta_{\text{top}} = \frac{P}{K_f} = \frac{13.0}{0.018 \times 10^6} = 0.72 \text{ mm}$$

This compares with $\Delta_{\text{top}} = 0.735$ mm for M2 (Table 4.5).

(5) Calculate the base moment in the wall.

$$M_{\text{wall}} = 403.2 \times \frac{14.0}{2} - 13.0 \times 14.0 = 2640 \text{ kN m}$$

This compares with $M_{\text{wall}} = 2156 + 468 = 2624$ kN m for M2 (Table 4.5).

The correlation here is good, mainly because the effect of the wall dominates the lateral stiffness.

6.7 CHECKING MODELS FOR PLATE BENDING

For checking, the plate bending problem being considered can be amended to a form for which a solution is available. This can be achieved by altering the boundary conditions and/or the loading. One should make an estimate of whether the checking model will be stiffer or more flexible than the finite element model. Corresponding results from the two models are then compared with some expectation of the sign of the difference between them.

A worthwhile approach is to use two checking models, one of which tends to be stiffer than the finite element model and the other tending to be more flexible. These should give results on either side of those being checked.

6.7.1 Formal solutions
The model can be converted to a form for which a formal solution is available — see for example Timoshenko & Woinowsky Krieger (1959) and Roark (1965).

6.7.2 Uniaxial bending models
It can be useful to take a strip of a plate in biaxial bending and treat it as being one-way spanning. This will tend to overestimate deformations and stresses. This is

particularly useful if the plate has a dominant span direction. For example, for the simply supported plate with side ratio of 2 : 1 and uniformly distributed load shown in Fig. 6.12(a), one can take a unit width strip as shown in Fig. 6.12(b) and treat it as a simply supported one-way span.

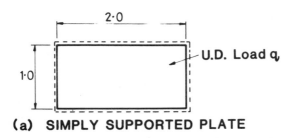

(a) SIMPLY SUPPORTED PLATE

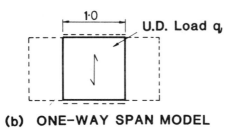

(b) ONE–WAY SPAN MODEL

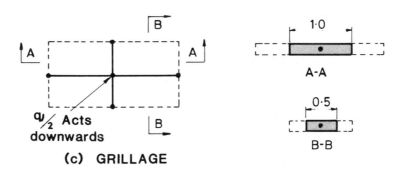

(c) GRILLAGE

Fig. 6.12 — Simply supported rectangular plate.

The maximum deflection and maximum moment for the strip can be calculated using:

$$\delta_{max} = \frac{5}{384} \frac{qa^4}{EI} \quad \text{and} \quad M_{max} = \frac{qa^2}{8}$$

where 'a' is the span of the strip.

The coefficients in the above expressions are quoted in Table 6.5. in which the corresponding coefficient from Timoshenko & Woinowsky-Krieger (1959) are also given.

6.7.3 Grillage models

Results for the simple grillage model shown in Fig. 6.12(c) are also shown in Table 6.5. For the grillage one quarter of the total load is applied as a point load at the centre. Member loads could also be used but it is easier to solve the point load case (the grillage of Fig. 6.12(c) can be solved by treating it as being singly statically indeterminate). The grillage gives a slightly better estimate of deflection but further overestimates the moments as compared with the one-way strip.

Fig. 6.13(a) shows a square simply supported uniformly distributed plate. Fig. 6.13(b) shows a simple grillage model. Again, one quarter of the total load is applied

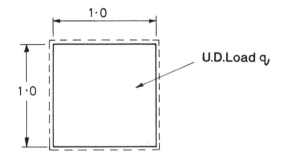

(a) SIMPLY SUPPORTED PLATE

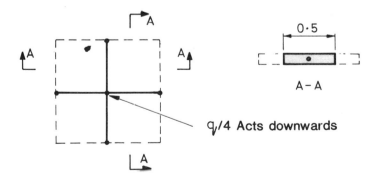

(b) GRILLAGE MODEL

Fig. 6.13 — Grillage model for checking square plate.

at the centre. Comparison of the maximum deflection and maximum moment values are quoted in Table 6.5.

Table 6.5 — Checking models for uniformly loaded simply supported plates

Problem	Model	Coefficient for δ_{max}	Coefficient for M_{max}
2a (rectangle, height a)	One-way strip	$0.01302 = \dfrac{5}{384}$	$0.125 = \dfrac{1}{8}$
	grillage	0.01225	0.1366
	Timoshenko	0.00922	0.1017
a (square, side a)	Grillage	$0.00251 = \dfrac{1}{192}$	$0.0625 = \dfrac{1}{16}$
	Timoshemko	0.00406	0.0479

$\delta_{max} = \text{coefficient} \times \dfrac{qa^4}{EI}$ $M_{max} = \text{coefficient} \times qa^2.$

The checking model results quoted in Table 6.5 do not give close correlation but do establish the likely orders of magnitude. Such information is valuable in checking.

6.8   CHECKING MODELS FOR ELASTIC STABILITY OF BUILDINGS

It is most important in structural design to guard against elastic instability of frame systems and insight into how such instability occurs is valuable when buckling load analysis is carried out.

In statically determinate systems buckling is triggered by a local instability but in indeterminate systems the instability effects may need to spread throughout the structure as the load increases to the critical value. For example, in a frame with moment connections, an unloaded braced column will have an end rotational stiffness of $4EI/L$. As the axial compressive load increases, the end stiffness decreases and can eventually become negative provided other elements which connect to the column can provide sufficient positive stiffness such that the joint stiffness is still positive. The negative stiffness spreads through the structure as the axial load is increased until the joint stiffnesses drop to zero and the frame buckles.

6.8.1   Sway mode buckling loads

Fig. 6.14 shows a simple model of sway buckling. A rigid column is pinned at its base and is supported laterally at the top by a spring of stiffness k. A displacement δ is imposed at the top. In the displaced position there is an overturning moment about the pin support $P\delta$ and a restoring moment from the spring $k\delta h$. For unstable equilibrium the restoring moment just matches the overturning moment, i.e.

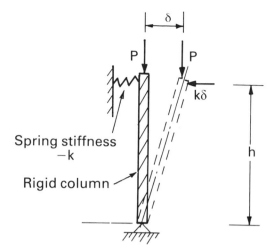

Fig. 6.14 — Sway buckling model.

$$P\delta = k\delta h$$

hence $P = P_{cr} = kh$ (6.25)

This is a useful model both for understanding the nature of a buckling situation and as a means of estimating sway critical loads. For example, consider the portal frame of Fig. 6.15(a). The total vertical load acting at beam level is P. A first approximation to the sway critical load is $P_{cr} = kh$ where k is the lateral stiffness at beam level. This is estimated by applying a lateral load W at beam level. Then $k = W/\delta$ where δ is the calculated deflection in the line of W. A similar calculation can be carried out for a pitched portal in which case the 'height' might be taken as less than that to the apex to be conservative — Fig. 6.15(b).

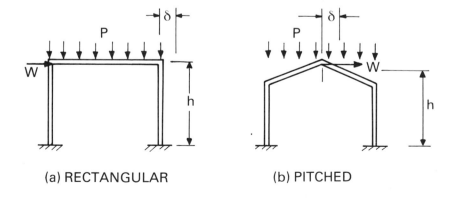

(a) RECTANGULAR   (b) PITCHED

Fig. 6.15 — Portal frames.

6.8.2 Multistorey frames

Shear mode frames

For rigid jointed (vierendeel) frames deforming in a shear mode, what is known as Horne's method uses a model close to that of equation (6.25). The frame is analysed under a uniformly distributed lateral load. At each storey, the storey height stiffness is calculated from

$$k = \frac{S}{\delta} \tag{6.26}$$

where S is the total shear at the top of the storey level, δ is the storey height 'drift'. The critical load at that storey level is then

$$P_{cr} = kh = \frac{S}{\delta} h \tag{6.27}$$

where h is the storey height. Horne (1975) recommends a reduction factor $=0.9$ on the value given by equation (6.27) to account for the fact that the column is not straight within the storey height. However, in view of other potential inaccuracies, this factor may be omitted. Horne also uses a lateral load on the frame equal to one hundredth of the vertical load but any intensity of lateral uniformly distributed load can be used provided the correct value of S is used in equation (6.27). The ratio P_{cr}/P, where P is the applied load, is calculated at each storey level to find the minimum value. In a checking situation the storey height drift can be calculated using equation (6.9) or equation (6.12).

Bending mode frames

For multistorey systems which deform predominantly in a bending mode a storey height sway mode is not a viable concept. Such frames suffer sway buckling in an overall mode and P_{cr} can be estimated as follows:

Assuming that

- the deflected shape of the building under sway is a straight line
- that the vertical load is uniformly distributed with height
- the lateral restoring load is also uniformly distributed with height.

Unstable equilibrium of the system of Fig. 6.16 will occur when:

overturning moment = restoring moment

i.e. $$\frac{P\Delta}{2} = W\frac{H}{2}$$

$$\therefore \quad P = P_{cr} = \frac{W}{\Delta}H = KH \tag{6.28}$$

where Δ is the top lateral deflection due to W. P is the applied vertical load, W is the integral of the internal shears which resist lateral movement and H is the total height. K can therefore be calculated as the top lateral stiffness based on a uniformly distributed lateral load using models described in section 6.6.

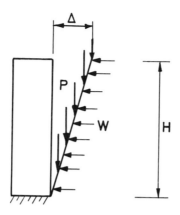

Fig. 6.16 — Sway stability of a building.

Equation (6.28) gives satisfactory estimates of buckling loads except where shear mode dominates (MacLeod & Marshall 1983).

Equation (6.28) can be used to estimate the sway buckling loads for a complete building.

7

Dynamic models

Problems with dynamic movements appear to be more prominent now than in the past probably because we can now model such behaviour more successfully. Typical situations where dynamic actions need to be considered are resonance, earthquake and blast loading.

In this chapter only a basic introduction to dynamic analysis is given. For more detailed information, see Clough & Penzien (1982).

7.1 THE EQUATIONS OF MOTION

7.1.1 Single degree of freedom systems

The basic mathematical models of dynamic behaviour are the equations of motion. The simplest form of these equations relates to the mass and undamped spring model of Fig. 7.1(a). It is:

$$M\frac{d^2u}{dt^2} + Ku = F(t) \tag{7.1}$$

where u is the displacement in the x-direction, K is the spring stiffness, M is the mass and $F(t)$ is the applied force or 'forcing function' which is a function of time t. Equation (7.1) can be solved for different forcing functions.

Some restraint to movement, i.e. damping, is always present and the commonest model for this is viscous damping which provides a restraining force proportional to velocity. Fig. 7.1(b) shows the system symbolically. The damper is a 'dashpot' and the equation of motion is:

$$M\frac{d^2u}{dt^2} + C\frac{du}{dt} + Ku = F(t) \tag{7.2}$$

where C is the damping constant.

Other types of damping such as frictional damping are used.

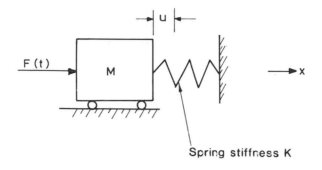

(a) MASS AND SPRING—NO DAMPING

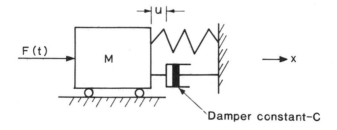

(b) WITH VISCOUS DAMPING

Fig. 7.1 — Single mass and spring models.

7.1.2 Multi degree of freedom systems

When the equations of motion relate to a multi degree of freedom system then equation (7.2) becomes:

$$\mathbf{M\ddot{u} + C\dot{u} + Ku = F}(t) \tag{7.3}$$

where the symbols represent the same attributes as for equation (7.2) but are now matrices.

$$\mathbf{\ddot{u}} = \frac{d^2\mathbf{u}}{dt^2} \qquad \mathbf{\dot{u}} = \frac{d\mathbf{u}}{dt}$$

7.1.3 Validity of the equations of motion

Equations of motion are equilibrium equations and the degree to which they can predict dynamic behaviour depends on the validity of the parameters K, C, M and $F(t)$.

Stiffness

The stiffness parameter K is a potential source of discrepancy. As discussed in Chapters 2 and 3, many of the assumptions used to establish K are highly approximate. For example, if one wished to model the dynamic behaviour of a reinforced concrete beam by treating it as a linear elastic material and using beam elements then such factors as cracking, variation in concrete quality, age of concrete, amount of steel, etc. could affect the result. Assumptions with respect to connection and support flexibility can also significantly affect the overall stiffness.

Mass

The mass of the structure itself can normally be modelled to good accuracy but the estimation of the live mass may be highly approximate. The mass matrix in equation (7.3) can either be 'lumped' or 'consistent'. The lumped form has only leading diagonal terms whereas the consistent mass matrix includes rotational and off-diagonal terms which model the linking effects of the masses. The consistent matrix is the more accurate but requires significantly more computing time to establish. In many cases the lumped matrix will give adequate accuracy.

Damping

Viscous damping is chosen more for ease of solution than for its relationship with the observed behaviour of structural systems. A frictional model would be more realistic but the viscous model can give acceptable results and fortunately natural frequencies are not highly affected by the degree of damping.

Damping is normally defined in terms of a percentage of critical damping (critical damping being the level at which no oscillation can develop when the system is set in free motion). Table A7 gives some typical damping coefficients.

Forcing function

If there is a forcing function it is often unlikely to be estimated accurately. Possibly the out of balance forces of rotating machinery can be estimated to reasonable accuracy but there is always a high degree of uncertainty in the estimation of blast forces and earthquake spectra.

Free vibration

When seeking natural frequencies there is a positive effect towards accuracy. The solution of equation (7.1) for the free vibration cases, i.e. with $F(t) = 0$, is

$$\omega = \sqrt{\frac{K}{M}} \tag{7.4}$$

where ω is the natural frequency in radians per second.

Taking the square root in equation (7.4) reduces the effect of errors in K and M on ω.

7.2 STRUCTURAL MODELS FOR DYNAMIC ANALYSIS

A few simple structural cases (mainly beam problems) can be solved by treating the properties of mass and stiffness as continuous functions. Normally however one uses a 'lumped' approach where stiffness and mass are defined at degrees of freedom as in equation (7.3). The question arises as to the best way of defining the stiffness and mass matrices for a given situation.

7.2.1 Condensed models

If there are N degrees of freedom in the structural model, then solution of equation (7.3) with $F(t) = 0$ will yield N natural frequencies and N mode shapes. When N is large, many of the natural frequencies tend to relate to unimportant modes and it is worthwhile to concentrate on that part of the structural behaviour which is relevant in the given situation.

This is achieved by condensing the stiffness and mass matrices. For example, Fig. 7.2(a) shows a three-storey plane frame model whose response to earthquake forces is to be investigated. A normal plane frame would have a total of twenty seven degrees of freedom and if all of these were used, a dynamic solution would produce twenty seven modes of vibration. For earthquake analysis one is normally only interested in a few of the higher modes corresponding to lateral deformation. The lateral dynamic behaviour is related mainly to the lateral freedoms and it is common to define the behaviour of the system in terms of only one lateral freedom at each floor level as illustrated in Fig. 7.2(b). The system is then visualised as a 'stick' model as in Fig. 7.2(c).

The procedure is therefore to choose a set of 'active' freedoms at which the stiffness and mass of the system are defined. In doing this the most important consideration is that the masses corresponding to the active freedoms should be those that dominate the dynamic response.

Condensed stiffness matrices

We wish to define the lateral stiffness of the frame of Fig. 7.2(a) in terms of the 'active' freedoms shown in Fig. 7.2(b); a process of condensation is defined as follows:

The structural stiffness matrix is partitioned in the form

$$\begin{bmatrix} \mathbf{K_{aa}} & \mathbf{K_{ap}} \\ \mathbf{K_{pa}} & \mathbf{K_{pp}} \end{bmatrix} \begin{Bmatrix} \Delta_a \\ \Delta_p \end{Bmatrix} = \begin{Bmatrix} \mathbf{P_a} \\ \mathbf{P_p} \end{Bmatrix} \tag{7.5}$$

where the subscripts **a** refer to active freedoms and **p** refer to passive freedoms (i.e. those freedoms which are not defined as active). No loads are applied at the passive freedoms and therefore $\mathbf{P_p} = 0$. Substituting this into the second row of equations (7.5) gives

$$\Delta_p = \mathbf{K_{pp}^{-1} K_{pa}} \Delta_a \tag{7.6}$$

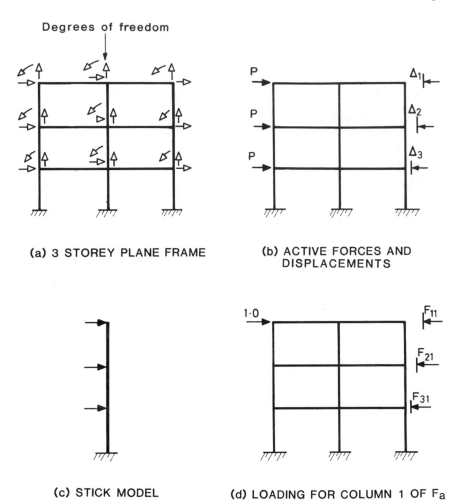

Fig. 7.2 — Condensed frame model.

Equation (7.6) is then substituted into the first row of equation (7.5) to give

$$[\mathbf{K_{aa}} - \mathbf{K_{ap}}\mathbf{K_{pp}^{-1}}\mathbf{K_{pa}}]\Delta_\mathbf{a} = \mathbf{P_a} \qquad (7.7)$$

Equation (7.7) is the condensed stiffness relationship. Different techniques are used to carry out this operation on the structural stiffness matrix.

To understand the concept of a condensed stiffness matrix it is worthwhile to consider a less formal method of obtaining it. This is illustrated using the frame of Fig. 7.2(a). Such a frame normally has twenty seven degrees of freedom as shown. We wish to define the stiffness in terms of the 3 lateral freedoms at each storey level — Fig. 7.2(b). A unit load is applied to the frame in the direction of master freedom 1 and the active deformations F_{11}, F_{21}, F_{31} are extracted from the results — Fig. 7.2(d).

The unit load is applied successively at all active freedoms. Each load case gives a vector of active deformations which are collected together in a flexiblity matrix which has the form

$$
\begin{Bmatrix} \Delta_1 \\ \Delta_2 \\ \Delta_3 \end{Bmatrix} = \begin{bmatrix} F_{11} & F_{12} & F_{13} \\ F_{21} & F_{22} & F_{23} \\ F_{31} & F_{32} & F_{33} \end{bmatrix} \begin{Bmatrix} P_1 \\ P_2 \\ P_3 \end{Bmatrix}
$$

i.e.

$$\Delta_a = F_a P_a \qquad\qquad (7.8)$$

This flexiblity relationship is inverted to give the stiffness relationship for the stick model:

$$P_a = F_a^{-1} \Delta_a = K_a \Delta_a \qquad\qquad (7.9)$$

where K_a is the condensed stiffness matrix which would be the same as that generated by use of equation (7.7).

Equation (7.9) is not an approximate version of the 21×21 stiffness matrix. It incorporates the effect of all freedoms shown in Fig. 7.2(a) and will therefore give the same results as would be obtained if all freedoms were included except that one would not be able to apply vertical or moment loads.

Equation (7.9) could be treated as a three-degree of freedom element stiffness relationship representing a three-storey frame element. This illustrates the arbitrariness of the choice of freedoms and elements in the stiffness method. An element is simply part (or all) of a model for which a force deformation relationship is available.

The process described above can be used for a set of frames and/or walls or for a complete building structure.

Condensed mass matrices
Condensed mass matrices can be created using a similar process to that defined by equation (7.7). In many cases a simple distribution of mass to the active nodes will suffice provided that the masses which correspond to the passive freedoms do not significantly affect the dynamic response.

7.2.2 Choice of master freedoms
One chooses as active freedoms those which dominate the vibration modes under consideration. If only the lower modes are needed they can often be represented by a small number of well chosen freedoms.

Single mass and spring system
A commonly used model is to reduce the complete system to a single mass and spring equivalent — Fig. 7.1. To do this a characteristic mass and a characteristic spring stiffness need to be defined.

Such a model will only be valid for systems that can be represented to a reasonable degree of approximation by a single degree of freedom. For example, a beam can be represented by a lateral freedom at mid-span, a cantilever by a lateral freedom at the tip and a slab panel by a lateral freedom at the centre of the panel.

Solutions for equations (7.1) or (7.2) are more readily obtained than for multi-degree of freedom systems and can sometimes be made available in the form of charts.

The single mass and spring model system can be useful as a checking model (section 6).

7.3 DYNAMIC SITUATIONS

7.3.1 Resonance

This situation occurs in the design of suspended floors, wind effects on slender structures, supports for rotating machinery, etc.

The basic problem is to identify the important natural frequencies of the system with a view to preventing any of them being close to the freqencies of the applied loading.

To estimate the natural frequency, equation (7.1) is solved without the forcing function. Taking out the forcing function simplifies the calculation. Neglecting damping is normally acceptable in such situations since the inaccuracy involved in doing this has a lesser effect on natural frequency than the inaccuracy involved in estimating the stiffness of the structure.

For natural frequency calculations there is no 'safe side' on which one can lean when estimating parameter values. One is looking for a critical natural frequency and overestimating a structural frequency has the same effect as underestimating it if the critical value is missed.

7.3.2 Earthquake response

The conventional approach to earthquake design has been to use a quasi-static approach to model the essentiallly dynamic effects of earthquake motions. With the availability of dynamic analysis software wide horizons of modelling have emerged and one would expect that dynamic analysis would become the norm in this application. To a degree this is happening but one should be careful not to overestimate the value of this approach as a general design tool.

The basic dynamic approach for earthquake situations is to set up a model of the structural system and to impose at its base a time-dependent earthquake input based on measurements of real earthquakes. Several methods of solving this problem are used.

In some cases where safety considerations are paramount, e.g. for nuclear installations, it may be necessary to try to ensure that the earthquake does not cause excursions into the plastic range. In such cases an elastic dynamic analysis can be performed to good advantage with a reasonable confidence that the results for a given earthquake input will give an acceptable prediction of behaviour. However, for conventional structures in areas of high earthquake risk, dynamic analysis indicates that the forces which would occur if the structure remains elastic are much greater than is conventionally used in the quasi-static approach. The conclusion is therefore made that during a strong motion earthquake, structures are likely to suffer plastic deformation. If this is so then plastic behaviour should be included in the model. This has been treated approximately by applying ductility factors to the elastic response but I am sceptical of the value of this approach.

A much more satisfactory method is to use a time history analysis where the dynamic behaviour is simulated as a set of time steps during which the material properties can vary. Another factor which has been shown to be most important is the effect of the soil stiffness on the structural response. Non-linear geometry effects and damping are also likely to be non-negligible.

Modelling for earthquake situations therefore presents the structural analyst with the ultimate challenge. In some cases one should take account of the soil, the structure, non-linear material behaviour, non-linear geometry effects and the non-linear time domain effects. Such models are used but are highly expensive and probably still give results to a very rough accuracy.

This is a topic of utmost importance in structural engineering. At the time of writing (December 1988) a major earthquake in Armenia has recently occurred in which modern buildings proved to be quite inadequate, causing many thousands of deaths. It is my impression that most advances in seismic design have been the result of observations of buildings which have been subject to severe earthquake conditions. Analytical modelling has an important place but its limitations should always be held in view.

For more detailed information on this subject see Green (1987), Clough & Penzien (1982) and Booth *et al.* (1988).

7.3.3 Blast loading
The effect of an airborne blast on a structure is that an initial pressure p_0 builds up in a negligibly short time and then diminishes to zero in time t. It is common to assume that the rate of decrease in pressure is linear — Fig. 7.3.

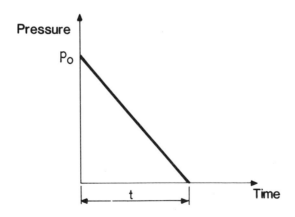

Fig 7.3   Blast loading model.

The response of the structure depends on the relative length of time t to the natural period of vibration of the structure T. If t is less than T then the structure does not have time to respond fully to the blast load, whereas if t is significantly greater than T the response is similar to a static load situation.

Local behaviour needs to be taken into account since parts of the structure can respond to the blast quite differently to the way the complete structure will respond.

The concept of treating the structure or part of the structure as a single mass and spring system is useful for blast loading. The spring will not remain elastic and an elasto-plastic relationship of the type illustrated in Fig. 2.7 is normally used. The ductility (section 2.3) of the spring is a most important parameter.

For further information on this subject see Alexander *et al.* (1970), ASCE (1985), Bulson (1985), Concrete Society (1984), Department of the Army (1969) and Biggs (1974).

8

Definitions and derivations

Notation
d   vector of element deformations
Δ   vector of deformations — normally system deformations
p   vector of element actions
P   vector of system loads
k   element stiffness matrix
K   structural stiffness matrix

This chapter gives derivations of some of the more important equations quoted elsewhere in the text and defines some terms which have general significance.

8.1   CONSTRAINTS

A constraint is in effect a compatibility condition imposed in a model. The term 'restraint' has the connotation of an imposed deformation — usually a support fixity. A 'constraint' links deformations relating to non-restrained freedoms. Constraints are normally viewed as being associated with rigid areas or with rigid connections in a model.

Fig. 8.1(a) shows a rigid beam system. At node 1 movements Δ_{y1} and θ_1 are defined. If the beam is rigid then the vertical movement at node 2 Δ_{y2} is related to Δ_{y1} and θ_1 by

$$\Delta_{y2} = \Delta_{y1} - L\theta_1 \tag{8.1}$$

Equation (8.1) is a typical constraint equation. The freedoms which relate to the deformations which define the movement of the rigid part are called 'master' freedoms and those relating to other deformations defined on the rigid part are 'slave' freedoms. In this case Δ_{y2} is a slave deformation and Δ_{y1} *and* θ_1 are master deformations.

Fig. 8.1(b) shows a rigid plane with a master node and a slave node. The master deformations corresponding to the three freedoms at the node 1 are Δ_{xm}, Δ_{ym}, and θ_m. The corresponding slave freedoms at node 2 are Δ_{xs}, Δ_{ys}, θ_s. These two sets of freedoms are related by:

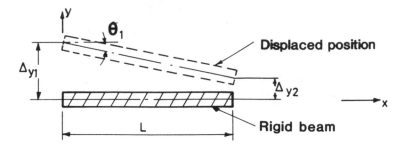

(a) LATERAL MOVEMENT OF RIGID BEAM

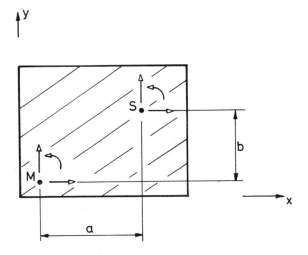

(b) MASTER AND SLAVE NODES ON A RIGID PLANE

Fig. 8.1 — Rigid body deformations.

$$\begin{Bmatrix} \Delta_{xs} \\ \Delta_{ys} \\ \theta_s \end{Bmatrix} = \begin{bmatrix} 1 & 0 & -b \\ 0 & 1 & a \\ 0 & 0 & 1 \end{bmatrix} \begin{Bmatrix} \Delta_{xm} \\ \Delta_{ym} \\ \theta_m \end{Bmatrix}$$

i.e.

$$\mathbf{d_s = A\ d_m} \qquad\qquad (8.2)$$

Equations (8.2) are the constraint equations for relating the movement of points in a rigid plane.

Constraints can be imposed using one of the following methods:

- By a Lagrange multiplier approach in which an extra freedom is added for each slave freedom (Gallacher 1975). The process of doing this is similar to adding further elements.

    A program which allows constraints to be imposed in this way should include:
    (a) The definition of all freedoms at a node as slaves to those at another node.
    (b) The definition of any single freedoms as a slave to a set of master freedoms. This would include the facility of simple equalization of freedoms.
    (c) The ability to introduce more complicated constraint relationships as data.
    (d) The ability to generate sets of constraints throughout the height or the width of a structure.

- By eliminating the slave freedoms by condensation of the structural stiffness matrix, (Gallacher 1975, MacLeod, Green, Wilson & Bhatt 1972). This method is not commonly used.

- By building them into the element stiffness matrix. This is illustrated for beam elements in section 8.5.1. Such rigid connections are also used with flat shell elements to simulate the effect of downstand beams — see section 4.4.4. Rigid connections can be added to most structural elements.

- Alternatively, constraints can be modelled using elements with high but finite stiffness. This has the advantage that the constraint equation facility is not needed but care must be taken not to give the rigid members too great a stiffness. A main cause of ill conditioning of the equations in a structural analysis is large differences between the terms of the structural stiffness matrix. I often use a rule that the bending stiffness of the 'rigid' elements should not be greater than 100 times the stiffness of adjacent elements.

8.2 DEGREE OF FREEDOM

'Action' is a generic term denoting forces (linear), moments (rotational), etc. For a structure the external actions are the applied loads and the internal actions are moments, shears, axial forces, etc., within the elements of the structure.

'Displacement' is a generic term denoting movements which are translational (linear), rotational, etc.

For each action there is a corresponding displacement. Such actions and displacements are vectors having position, direction and magnitude. A 'degree of freedom' (or 'freedom') is the position and direction which is common to an action and its corresponding displacement. A straight arrow is normally used to denote a translational freedom and a curved arrow denotes a rotational freedom — Fig. 8.2.

Other types of freedom which correspond to cross derivatives and other higher order derivatives of displacement are used in some finite element formulations.

8.3 BOUND THEOREMS

8.3.1 Lower bound theorem

This is a most important concept. The theorem can be stated as follows:

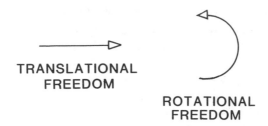

Fig. 8.2 — Degrees of freedom.

If a set of internal forces is identified which is in equilibrium with the applied load on a structure and the yield criterion is not violated then the corresponding applied load is less than or equal to the collapse load.

The 'yield criterion' is normally stated as 'the yield stress is not anywhere exceeded' but the condition should also include the requirement that the ductility is everywhere adequate to sustain yield stresses.

A consequence of the lower bound theorem is that if the conditions are satisfied, then use of the internal forces to size the members of the structure will result in a safe design. This is because the real collapse load must be greater than the one for which the structure has been designed.

Designers often use the lower bound theorem implicitly when they choose a set of forces on which to base their member sizing calculations.

To illustrate the behaviour of a system in relation to the lower bound theorem consider the rigid block shown in Fig. 8.3(a). The block is supported on two springs and is restrained from rotation. A load W is applied in the direction shown.

The spring stiffnesses are denoted by k_1 and k_2 and the spring loads are P_1 and P_2. The elastic load distribution would be:

$$P_1 = \frac{k_1}{k_1 + k_2}\, W, \qquad P_2 = \frac{k_2}{k_1 + k_2}\, W \qquad\qquad (8.3)$$

Defining $\beta = k_1/k_2$ gives:

$$P_1 = \frac{\beta}{1 + \beta}\, W, \qquad P_2 = \frac{1}{1 + \beta}\, W \qquad\qquad (8.4)$$

and

$$P_1/P_2 = \beta \qquad\qquad (8.5)$$

Suppose instead of the value of β defined by equation (8.5) an arbitrary value of the spring load ratio 'α' is used for sizing the springs. Spring 2 will then be designed to yield at a load P_y and spring 1 to yield at a load of αP_y. The collapse load W_c would then be

$$W_c = \alpha P_y + P_y = (1 + \alpha)P_y \qquad\qquad (8.6)$$

Fig. 8.3(b) shows the load deflection diagram for the system. At a load of W_1 spring 1

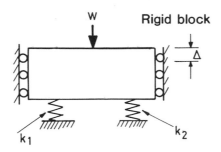

(a) RIGID BLOCK ON TWO SPRINGS

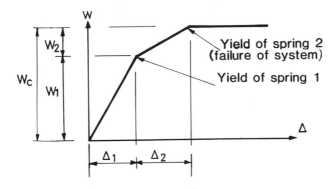

(b) LOAD DEFLECTION DIAGRAM

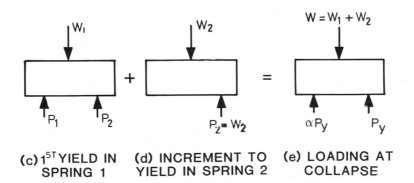

(c) 1ST YIELD IN SPRING 1   **(d) INCREMENT TO YIELD IN SPRING 2**   **(e) LOADING AT COLLAPSE**

Fig. 8.3 — Rigid block of yielding springs.

yields (spring 1 will yield first if $\alpha < \beta$) at deflection Δ_1. A further increment of load, W_2, is needed to reach yield in spring 2 with a corresponding increase in deflection Δ_2.

Fig. 8.3(c) shows the forces in the block at yield in spring 1.
At this condition:

$$P_1 = \alpha P_y \qquad P_2 = P_1/\beta = \frac{\alpha}{\beta} P_y$$

$$W_1 = P_1 + P_2 = \alpha P_y + \frac{\alpha}{\beta} P_y = \alpha\left(1 \times \frac{1}{\beta}\right) P_y$$

and

$$\Delta_1 = P_1/k_1 = \alpha P_y/k_1$$

During the increment of load W_2, P_1 stays constant (at αP_y) and spring 1 provides no restraint to movement. This situation is illustrated in Fig. 8.3(d). At yield in spring 2 the total applied load is $W_1 + W_2$, $P_1 = \alpha P_y$ and $P_2 = P_y$. Therefore for equilibrium

$$W_1 + W_2 = (1 + \alpha)P_y$$

therefore

$$W_2 = (1 + \alpha)P_y - \alpha\left(1 + \frac{1}{\beta}\right)P_y = \left(1 - \frac{\alpha}{\beta}\right)P_y$$

and

$$\Delta_2 = W_2/k_2 = \left(1 - \frac{\alpha}{\beta}\right)\frac{P_y}{k_2}$$

A ductility requirement factor — drf — is defined here as the ratio of deformation at yield of the system to deformation at first yield, i.e.

$$\text{drf} = \frac{\Delta_1 + \Delta_2}{\Delta_1} = \frac{\dfrac{\alpha P_y}{k_1} + \left(1 - \dfrac{\alpha}{\beta}\right)\dfrac{P_y}{k_2}}{\dfrac{\alpha P_y}{k_1}} = \frac{\beta}{\alpha} \tag{8.7}$$

Therefore if spring 1 is to maintain its yield load up to yield in spring 2 it needs a ductility factor not less than the ratio β/α.

If $\beta < \alpha$, then spring 2 will yield first and drf $= \alpha/\beta$. In general, the less accurate the estimate of distribution of load, the greater will be the ductility requirement and in this case these factors are directly proportional.

In most cases the ductility will be sufficient but yielding may be associated with cracking and should be avoided. Thus to maintain serviceability it is good design policy to keep the assumed distribution of load as close to the elastic distribution as possible and then provide details which will allow ductility to cater for any variation from this.

I am somewhat wary of the lower bound theorem. It can be used to neglect structural actions (such as torsion) and it seems possible that such neglect could, in certain circumstances, prevent the assumed plastic state from being realized.

8.3.2　Upper bound theorem

The statement of this theorem is:

> If a set of internal forces is identified which gives a collapse mechanism then the corresponding applied load is greater than or equal to the true collapse load.

This theorem is important when using the limit analysis approach to finding a collapse load but is seldom used in computer modelling where the load is stepped up to a collapse condition.

8.4　THE PRINCIPLE OF CONTRAGREDIENCE

By means of this principle (Livesley 1975) a link is identified between compatibility and equilibrium which is useful when dealing with element stiffness relationships.

Consider the rigid beam shown in Fig. 8.4. Freedoms 1 and 2 can be considered as master freedoms for lateral movement and freedom 3 as a slave freedom. The displacements and corresponding forces are Δ_1, Δ_2, Δ_3 and P_1, P_2, P_3 respectively.

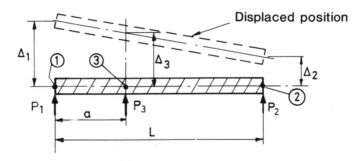

Fig. 8.4 — Rigid beam with lateral forces.

The compatibility condition is:

$$\Delta_3 = \left[\left(1 - \frac{a}{L} \right) \frac{a}{L} \right] \left\{ \begin{array}{c} \Delta_1 \\ \Delta_2 \end{array} \right\}$$

i.e. $\Delta_s = \mathbf{A}\,\Delta_m$　　　　　　　　　　　　　　　　　　　　　　　　　　(8.8)

where the subscript 's' denotes a slave quantity and 'm' denotes a master quantity.

The equilibrium relationship (for P_3 to be statically equivalent to P_1 and P_2) is:

$$\left\{ \begin{array}{c} P_1 \\ P_2 \end{array} \right\} = \left[\begin{array}{c} \left(1 - \frac{a}{L} \right) \\ \frac{a}{L} \end{array} \right] P_3$$

i.e. $\mathbf{P_m} = \mathbf{B} \ \mathbf{P_s}$

Note that $\mathbf{B} = \mathbf{A^T}$

We wish to show the generality of this relationship.

$\mathbf{P_m}$ and $\mathbf{P_s}$ form an equilibrium set and $\mathbf{\Delta_m}$ and $\mathbf{\Delta_s}$ form a corresponding compatible set. Therefore the principle of virtual work can be invoked, i.e.

$$\mathbf{\Delta_m^T} \ \mathbf{P_m} = \mathbf{\Delta_s^T} \ \mathbf{P_s} \tag{8.9}$$

From equation (8.8)

$$\mathbf{\Delta_s^T} = \mathbf{\Delta_m^T} \ \mathbf{A^T} \tag{8.10}$$

Substituting equation (8.10) into equation (8.9) gives

$$\mathbf{\Delta_m^T} \ \mathbf{P_m} = \mathbf{\Delta_m^T} \ \mathbf{A^T P_s}$$

hence

$$\mathbf{P_m} = \mathbf{A^T P_s} \tag{8.11}$$

Therefore

$$\mathbf{B} = \mathbf{A^T}$$

Note that corresponding quantities appear on opposite sides of the equations, i.e.

$$\mathbf{\Delta_s} = \mathbf{A} \ \mathbf{\Delta_m}$$

$$\mathbf{P_m} = \mathbf{A^T} \ \mathbf{P_s}$$

This correspondence between equilibrium and compatibility is called the 'principle of contragredience'. Before being introduced to this concept I thought of equilibrium and compatibility as being separate criteria. They are, however, linked by the principle of virtual work.

It is common to observe that when analysing indeterminate structures three basic sets of equations need to be stated, i.e. force-deformation, equilibrium and compatibility. Using the principle of contragredience the number of independent sets is effectively reduced to two.

8.5   METHODS FOR DEVELOPING ELEMENT STIFFNESS MATRICES

The basic assumptions made in establishing element stiffness matrices are discussed in section 3.1.1.

In this section means of formulating such matrices on the basis of these assumptions are described.

A common classification of methods for finite element derivation is minimum potential energy and weighted residual methods. The former requires the problem to be stated in the form of a 'functional' which is an integral defining the energy potential of the system. The first variation of this is used to establish the stiffness matrix.

For the weighted residual approach the problem is stated in the form of a governing differential equation with boundary conditions; that virtually any system which can be defined in this way can be solved using the finite element method was a major discovery.

It is conventional to define structural problems using the principle of minimum potential energy although the weighted residual method can also be used giving the same results.

In fact, structural element stiffness matrices can generally be formulated using the principle of virtual work. Whether this is a special case of the principle of minimum potential energy or vice versa is debatable. The principle of virtual work has a physical interpretation which I think I understand better than the potential energy approach. Therefore virtual work is used in the derivations given here.

8.5.1   Uniform beam elements
Beam elements with uniform cross-section and end loading are type A elements which do not need mesh refinement — see section 3. The approach used here for element derivation breaks down the problem into basic parts and shows how these can be assembled to form complete element relationships. I find this method helpful for basic understanding and it is useful for developing specialized elements of this type.

Bending terms
Firstly the stiffness in terms of a self-straining system is defined. A 'self straining' system is one in which rigid body movements are not considered. The deformed shape of the element is fully defined by the self-straining systems. It therefore contains useful information about behaviour of an element. It is interesting to note that the first published element derivation of the modern 'finite element' type used a self-straining system to define the displacement field (Argyris 1954).

For bending, Fig. 8.5(a) shows a simply supported beam with end moments m_{A1} and m_{A2}. This is a suitable self-straining system. (A cantilever with end shear and moment would also form a suitable system for this purpose.) The system of Fig. 8.5(a) is denoted as system A.

The flexibility relationship for this system is:

$$\left\{ \begin{array}{c} \theta_{A1} \\ \\ \theta_{A2} \end{array} \right\} = \frac{L}{EI} \begin{bmatrix} \dfrac{1}{3} & -\dfrac{1}{6} \\ \\ -\dfrac{1}{6} & \dfrac{1}{3} \end{bmatrix} \left\{ \begin{array}{c} m_{A1} \\ \\ m_{A2} \end{array} \right\} \tag{8.12}$$

i.e

$$\mathbf{d}_{ss} = \mathbf{f}_{ss}\mathbf{p}_{ss}$$

This subscript 'ss' stands for self-straining.

The symbols used in equation (8.12) are defined in Fig. 8.5(a). The terms of the matrix can be derived using the principle of virtual work.

Equation (8.12) is a self-straining relationship which fully defines the bending characteristics of the uniform beam element under end loading. The inverse form of equation (8.12) is:

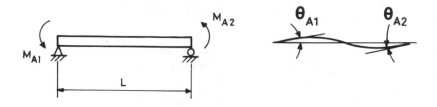

(a) SELF-STRAINING SYSTEM FOR A BENDING ELEMENT-SYSTEM A

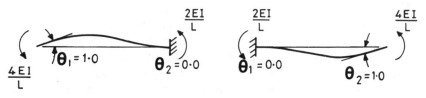

End moments for column 1   End moments for column 2

(b) END MOMENTS FOR k_{ss} (Equation (8·13))

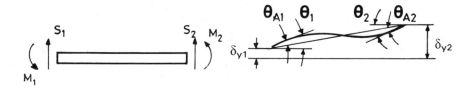

(c) GENERAL FREEDOMS FOR A BENDING ELEMENT – SYSTEM B

Fig. 8.5 — Uniform bending element.

$$\begin{Bmatrix} m_{A1} \\ m_{A2} \end{Bmatrix} = \frac{EI}{L} \begin{bmatrix} 4 & 2 \\ 2 & 4 \end{bmatrix} \begin{Bmatrix} \theta_{A1} \\ \theta_{A2} \end{Bmatrix} \tag{8.13}$$

i.e.

$$\mathbf{p}_{ss} = \mathbf{k}_{ss}\mathbf{d}_{ss}$$

These are well known as part of the traditional slope deflection equations.

To establish the complete bending stiffness matrix defined in equation (8.17) only the conditions of equilibrium and compatibility are required. No further force deformation information is needed.

$\mathbf{k}_{ss}$ is a basic element stiffness matrix. It is worthwhile to consider what the terms of a stiffness matrix mean. I like to look at the stiffness matrix column by column. The first column of $\mathbf{k}_{ss}$ gives the moments that are required to hold the beams in a displaced shape with $\theta_{A1} = 1$ and $\theta_{A2} = 0$ — see Fig. 8.5(b). That is, column i of the matrix is a list of the end forces which will hold the deformed shape $\theta_i = 1.0$ with all other nodal deformations at zero. This view of the terms of a stiffness matrix links with the shape function approach for deriving them as described in section 8.5.3.

Addition of end shears

In Fig. 8.5(c), system B, the end shears are added to the moments of Fig. 8.5(a). We want to relate the forces of system B to those of system A. Equilibrium conditions give:

$$\begin{Bmatrix} M_1 \\ S_1 \\ M_2 \\ S_2 \end{Bmatrix} = \begin{bmatrix} 1 & 0 \\ \dfrac{1}{L} & \dfrac{1}{L} \\ 0 & 1 \\ -\dfrac{1}{L} & -\dfrac{1}{L} \end{bmatrix} \begin{Bmatrix} m_{A1} \\ m_{A2} \end{Bmatrix}$$

i.e.

$$\mathbf{p}_{bz} = \mathbf{H}^T \, \mathbf{p}_{ss} \qquad (8.14)$$

where the subscript 'bz' stands for bending in the z-plane.

By the principle of contragredience (section 8.4)

$$\mathbf{d}_{ss} = \mathbf{H} \, \mathbf{d}_{bz} \qquad (8.15)$$

Substituting equations (8.14) and (8.15) into equation (8.13) gives

$$\mathbf{p}_{bz} = \mathbf{H}^T \, \mathbf{k}_{ss} \, \mathbf{H} \, \mathbf{d}_{bz} \qquad (8.16)$$

Performing the transformation $\mathbf{H}^T \, \mathbf{k}_{ss} \, \mathbf{H}$ gives:

$$\begin{Bmatrix} M_1 \\ S_1 \\ M_2 \\ S_2 \end{Bmatrix} = EI_z \begin{bmatrix} \dfrac{4}{L} & \dfrac{6}{L^2} & \dfrac{2}{L} & -\dfrac{6}{L^2} \\ \dfrac{6}{L^2} & \dfrac{12}{L^3} & \dfrac{6}{L^2} & -\dfrac{12}{L^3} \\ \dfrac{2}{L} & \dfrac{6}{L^2} & \dfrac{4}{L} & -\dfrac{6}{L^2} \\ -\dfrac{6}{L^2} & -\dfrac{12}{L^3} & -\dfrac{6}{L^2} & \dfrac{12}{L^3} \end{bmatrix} \begin{Bmatrix} \theta_1 \\ \delta_{y1} \\ \theta_2 \\ \delta_{y2} \end{Bmatrix} \qquad (8.17)$$

i.e.

$$\mathbf{p}_{bz} = \mathbf{k}_{bz}\,\mathbf{d}_{bz}$$

The transformation of $\mathbf{k}_{ss}$ into $\mathbf{k}_{bz}$ as described above is a standard approach for this type of element. Similar transformations are carried out to rotate the element or to add rigid ends as described later.

$\mathbf{k}_{by}$ (the bending matrix in the y-plane) is the same as $\mathbf{k}_{bz}$ except that $I = I_y$.

Axial terms

Fig. 8.6(a) shows a self-straining system for a bar element. Fig. 8.6(b) shows the more general system of freedoms. A similar process to that for the bending element gives the stiffness relationship:

$$\begin{Bmatrix} N_1 \\ N_2 \end{Bmatrix} = \frac{EA}{L}\begin{bmatrix} 1 & -1 \\ -1 & 1 \end{bmatrix}\begin{Bmatrix} \delta_{x1} \\ \delta_{x2} \end{Bmatrix} \tag{8.18}$$

i.e.

$$\mathbf{p}_n = \mathbf{k}_n\,\mathbf{d}_n$$

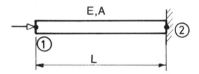

(a) SELF STRAINING SYSTEM FOR A BAR ELEMENT

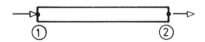

(b) GENERAL FREEDOMS FOR A BAR ELEMENT

Fig. 8.6 — Uniform bar element.

Plane frame bending element

Equations (8.17) and (8.18) are combined to give the basic plane frame element stiffness matrix relative to the freedoms shown in Fig. 3.12(a).

$$
\begin{Bmatrix} M_1 \\ S_1 \\ N_1 \\ M_2 \\ S_2 \\ N_2 \end{Bmatrix}
=
\begin{bmatrix}
\dfrac{4EI_z}{L} & \dfrac{6EI_z}{L^2} & 0 & \dfrac{2EI_z}{L} & -\dfrac{6EI_z}{L^2} & 0 \\[2mm]
\dfrac{6EI_z}{L^2} & \dfrac{12EI_z}{L^3} & 0 & \dfrac{6EI_z}{L^2} & -\dfrac{12EI_z}{L^3} & 0 \\[2mm]
0 & 0 & \dfrac{EA}{L} & 0 & 0 & -\dfrac{EA}{L} \\[2mm]
\dfrac{2EI_z}{L} & \dfrac{6EI}{L^2} & 0 & \dfrac{4EI_z}{L} & -\dfrac{6EI_z}{L^2} & 0 \\[2mm]
-\dfrac{6EI_z}{L^2} & -\dfrac{12EI_z}{L^3} & 0 & -\dfrac{6EI_z}{L^2} & \dfrac{12EI_z}{L^3} & 0 \\[2mm]
0 & 0 & -\dfrac{EA}{L} & 0 & 0 & \dfrac{EA}{L}
\end{bmatrix}
\begin{Bmatrix} \theta_1 \\ \delta_{y1} \\ \delta_{x1} \\ \theta_2 \\ \delta_{y2} \\ \delta_{x2} \end{Bmatrix}
\tag{8.19}
$$

$$\mathbf{p_{pf}} = \mathbf{k_{pf}}\,\mathbf{d_{pf}}$$

where subscript 'pf' stands for 'plane frame'.

St Venant torsion

The St Venant torsion element is shown in Fig. 3.10(a). Its form is identical to that of the axial force element and its stiffness matrix is:

$$
\begin{Bmatrix} T_1 \\ T_2 \end{Bmatrix} = \frac{CG}{L} \begin{bmatrix} 1 & -1 \\ -1 & 1 \end{bmatrix} \begin{Bmatrix} \varphi_1 \\ \varphi_2 \end{Bmatrix}
\tag{8.20}
$$

i.e.

$$\mathbf{p_t} = \mathbf{k_t}\,\mathbf{d_t}$$

Space frame element

The space frame element normally has six degrees of freedom at each end and is a combination of the bar element matrix $\mathbf{k_n}$, the z-axis bending matrix $\mathbf{k_{bz}}$, the y-axis bending matrix $\mathbf{k_{by}}$ (which has the same form as $\mathbf{k_{bz}}$) and the St Venant torsion matrix $\mathbf{k_t}$ — Fig. 3.12(c). If geometric non-linearity is neglected then these four elements are assumed to be uncoupled. The uncoupling is best understood if the complete matrix is written in the partitioned form:

$$
\begin{Bmatrix} \mathbf{p_n} \\ \mathbf{p_{bz}} \\ \mathbf{p_{by}} \\ \mathbf{p_t} \end{Bmatrix}
=
\begin{bmatrix}
\mathbf{k_n} & 0 & 0 & 0 \\
0 & \mathbf{k_{bz}} & 0 & 0 \\
0 & 0 & \mathbf{k_{by}} & 0 \\
0 & 0 & 0 & \mathbf{k_t}
\end{bmatrix}
\begin{Bmatrix} \mathbf{d_n} \\ \mathbf{d_{bz}} \\ \mathbf{d_{by}} \\ \mathbf{d_t} \end{Bmatrix}
\tag{8.21}
$$

i.e.

$$\mathbf{p}_{sf} = \mathbf{k}_{sf}\, \mathbf{d}_{sf}$$

where the subscript 'sf' stands for 'space frame'.

For programming purposes, $\mathbf{k}_{sf}$ is normally partitioned in relation to the freedoms at ends 1 and 2, i.e. in the form:

$$\begin{Bmatrix} \mathbf{p_1} \\ \mathbf{p_2} \end{Bmatrix} = \begin{bmatrix} \mathbf{k_{11}} & \mathbf{k_{12}} \\ \mathbf{k_{21}} & \mathbf{k_{22}} \end{bmatrix} \begin{Bmatrix} \mathbf{d_1} \\ \mathbf{d_2} \end{Bmatrix} \tag{8.22}$$

Shear deformation

Having followed through the derivation to this stage, we backtrack to the self-straining bending element — Fig. 8.5(a) — to show how shear deformation can be added.

Shear deformation causes the end rotations to increase. Under a moment m_{A1} at end 1 the shear is constant over the length with a value of m_{A1}/L. Fig. 8.7 shows the deformed shape of the element. The rotations can be calculated using the principle of virtual work giving $\theta_{s1} = \theta_{s2} = m_{A1}/(L\overline{A}G)$. The rotations due to m_{A2} have the value $\theta_{s1} = \theta_{s2} = m_{A2}/(L\overline{A}G)$.

The relationship given as equation (8.12) is therefore modifed by shear deformation to:

$$\begin{Bmatrix} \theta_{A1} \\ \\ \\ \\ \theta_{A2} \end{Bmatrix} = \begin{bmatrix} \left(\dfrac{L}{3EI} + \dfrac{L}{L\overline{A}G} \right) & \left(-\dfrac{L}{6EI} + \dfrac{1}{L\overline{A}G} \right) \\ \\ \left(-\dfrac{L}{6EI} + \dfrac{L}{L\overline{A}G} \right) & \left(\dfrac{L}{3EI} + \dfrac{1}{L\overline{A}G} \right) \end{bmatrix} \begin{Bmatrix} m_{A1} \\ \\ \\ \\ m_{A2} \end{Bmatrix} \tag{8.23}$$

Fig. 8.7(b) shows the rotations at end 1. θ_{A1} is now an average rotation of the end of the member (the section is no longer plane). It has a bending component (du/dx) and a shear component θ_{s1}.

Equation (8.23) can be inverted to give the corresponding stiffness matrix. The transformation to add the end shears (as in equation (8.16)) is then applied to produce a modified form of $\mathbf{k_{bz}}$. $\mathbf{k_{by}}$ can be similarly modified.

Rigid ends

We wish to define the stiffness characteristics of the plane frame element shown in Fig. 8.8(a). It has a flexible part ab and rigid ends with nodes 1 and 2. The element freedoms are at nodes 1 and 2 as shown.

The stiffness matrix of the flexible part is defined by the plane frame relationship of equation (8.19).

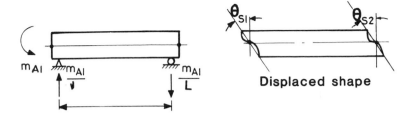

(a) SHEAR DEFORMATION OF BEAM DUE TO m_{AI}

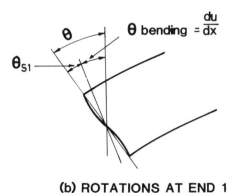

(b) ROTATIONS AT END 1

Fig. 8.7 — Rotations due to shear deformation.

The relationship between the end actions on the flexible part — $\mathbf{p_{pf}}$ and those at nodes 1 and 2 — $\mathbf{p_{12}}$ (Fig. 8.8(b)) is:

$$
\begin{Bmatrix} M_{z1} \\ p_{y1} \\ p_{x1} \\ M_{z2} \\ p_{y2} \\ p_{x2} \end{Bmatrix} = \begin{bmatrix} 1 & X_1 & Y_1 & 0 & 0 & 0 \\ 0 & 1 & 0 & 0 & 0 & 0 \\ 0 & 0 & 1 & 0 & 0 & 0 \\ 0 & 0 & 0 & 1 & -X_2 & Y_2 \\ 0 & 0 & 0 & 0 & 1 & 0 \\ 0 & 0 & 0 & 0 & 0 & 1 \end{bmatrix} \begin{Bmatrix} M_1 \\ S_1 \\ N_1 \\ M_2 \\ S_2 \\ N_2 \end{Bmatrix} \qquad (8.24)
$$

i.e.

$$\mathbf{p_{12}} = \mathbf{T^T}\, \mathbf{p_{pf}}$$

The corresponding relationship between the deformations is:

$$\mathbf{d_{pf}} = \mathbf{T}\, \mathbf{d_{12}} \qquad (8.25)$$

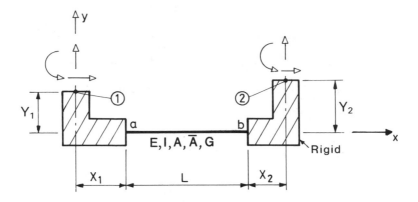

(a) ELEMENT GEOMETRY AND FREEDOMS

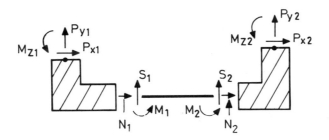

(b) MEMBER ACTIONS

Fig. 8.8 — Beam element with rigid ends.

Substituting equations (8.25) and (8.24) into equation (8.19) gives:

$$\mathbf{p}_{12} = \mathbf{T}^T \, \mathbf{k}_{pf} \, \mathbf{T} \, \mathbf{d}_{12} \qquad (8.26)$$

This is the stiffness relationship for the plane frame element with rigid ends.

Rotation of coordinate axes

It is normal to define the element properties in relation to a global coordinate system and therefore the forces and deformations need to be rotated from the member directions. The procedure is illustrated for the plane frame element of Fig. 3.12(a).

If the local axes are x and y, the global axes are x', y' and the angle between the axes is θ as defined in Fig. 8.9 then the relationship between the member forces in the two coordinate systems is:

$$\left\{\begin{array}{c} M'_1 \\ p'_{y1} \\ p'_{x1} \\ M'_2 \\ p'_{y2} \\ p'_{x2} \end{array}\right\} = \left[\begin{array}{cccccc} 1 & 0 & 0 & 0 & 0 & 0 \\ 0 & \cos\theta & \sin\theta & 0 & 0 & 0 \\ 0 & -\sin\theta & \cos\theta & 0 & 0 & 0 \\ 0 & 0 & 0 & 1 & 0 & 0 \\ 0 & 0 & 0 & 0 & \cos\theta & \sin\theta \\ 0 & 0 & 0 & 0 & -\sin\theta & \cos\theta \end{array}\right] \left\{\begin{array}{c} M_1 \\ S_1 \\ N_1 \\ M_2 \\ S_2 \\ N_2 \end{array}\right\} \qquad (8.27)$$

i.e.

$$\mathbf{p}'_{pf} = \mathbf{R}^T \, \mathbf{p}_{pf}$$

hence

$$\mathbf{d}_{pf} = \mathbf{R} \, \mathbf{d}'_{pf} \qquad (8.28)$$

Substituting equations (8.28) and (8.27) into equation (8.19) gives

$$\mathbf{p}'_{pf} = \mathbf{R}^T \, \mathbf{k}_{pf} \, \mathbf{R} \, \mathbf{d}'_{pf} \qquad (8.29)$$

This type of transformation is readily extended to three dimensions.

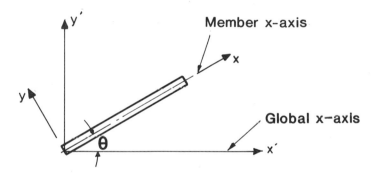

Fig. 8.9 — Member and global axes.

Summary

In this section some techniques used in the development of matrices for type A elements are described. The self-straining form of a stiffness relationship reduces the force deformation characteristics to a basic level which, hopefully, promotes better understanding of the assumptions.

8.5.2   The displacement function method

This method is used for type B elements but so that the size of the matrices involved is kept to a minimum, the process is demonstrated using a uniform bar element.

The assumptions for this element are discussed in section 3.3. Fig. 3.5 shows the element.

The steps in the process are as follows:

Define the nodal displacements **d** and nodal forces **p** corresponding to the degrees of freedom, i.e.

$$\mathbf{d} = \begin{Bmatrix} u_1 \\ u_2 \end{Bmatrix} \quad \mathbf{p} = \begin{Bmatrix} n_1 \\ n_2 \end{Bmatrix} \tag{8.30}$$

Choose a displacement function. In this case:

$$\mathbf{u} = \alpha_1 + \alpha_2 x = [1 \ x] \begin{Bmatrix} \alpha_1 \\ \alpha_2 \end{Bmatrix} \tag{8.31}$$

i.e.

$$\mathbf{u} = \mathbf{F} \, \alpha$$

where **α** is the vector of coefficients.

Express **d** in terms of **α**

$$\begin{Bmatrix} u_1 \\ u_2 \end{Bmatrix} = \begin{bmatrix} 1 & 0 \\ 1 & L \end{bmatrix} \begin{Bmatrix} \alpha_1 \\ \alpha_2 \end{Bmatrix} \quad \begin{array}{l} \text{substitute } x = 0 \text{ in equation (8.31)} \\ \text{substitute } x = L \text{ in equation (8.31)} \end{array}$$

i.e.

$$\mathbf{d} = \mathbf{A} \, \alpha \tag{8.32}$$

Invert equation (8.32) to get:

$$\begin{Bmatrix} \alpha_1 \\ \alpha_2 \end{Bmatrix} = \begin{bmatrix} 1 & 0 \\ -\dfrac{1}{L} & \dfrac{1}{L} \end{bmatrix} \begin{Bmatrix} u_1 \\ u_2 \end{Bmatrix} \tag{8.33}$$

i.e.

$$\alpha = \mathbf{A}^{-1} \mathbf{d}$$

Express **ε** the strain vector in terms of **α**. In this case the only strain component that needs to be considered is ε_x. In most other situtations there is more than one component so we treat it as a vector.

$$\varepsilon_x = \delta u / \delta x = [0 \quad 1] \alpha$$

i.e.

$$\varepsilon = \mathbf{Q} \, \alpha \tag{8.34}$$

Substitute equation (8.33) into equation (8.34)

$$\varepsilon = \mathbf{Q} \, \mathbf{A}^{-1} \mathbf{d}$$

i.e.

$$\varepsilon = \mathbf{B} \mathbf{d} \tag{8.35}$$

where

$$\mathbf{B} = \mathbf{Q} \, \mathbf{A}^{-1}$$

Define the constitutive relationship

$$\sigma_x = E\varepsilon_x, \quad \text{i.e. } \boldsymbol{\sigma} = \mathbf{D} \, \boldsymbol{\varepsilon} \tag{8.36}$$

Now use the principle of virtual work to establish the element stiffness. We have an equilibrium set, i.e. the external actions **p** and the internal stress **σ**. We also have a corresponding compatible set, i.e. the external deformations **d** and the internal strains **ε**. The principle of virtual work is now applied to these sets.

Note that the external work done by forces **p** through corresponding deformations **d** is given by the scalar product:

$$\{u_1 \ u_2\} \begin{Bmatrix} n_1 \\ n_2 \end{Bmatrix} = \mathbf{d}^{\mathrm{T}} \, \mathbf{p}$$

The internal work done by σ_x acting on area $dydz$ through the deformation $\varepsilon_x dx$ is:
Internal work = $\int \varepsilon_x dx \, \sigma_x dydz = \int \varepsilon_x \sigma_x \, dvol$
Therefore the total internal work is given by:

Internal work = $\displaystyle\int \boldsymbol{\varepsilon}^{\mathrm{T}} \, \boldsymbol{\sigma} \, dvol$

External work = Internal work
i.e.

$$\mathbf{d}^{\mathrm{T}} \, \mathbf{p} = \int \boldsymbol{\varepsilon}^{\mathrm{T}} \, \boldsymbol{\sigma} \, dvol \tag{8.37}$$

Substitute equations (8.35) and (8.36) into equation (8.37) to give:

$$\mathbf{d}^{\mathrm{T}}\mathbf{p} = \mathbf{d}^{\mathrm{T}} \int \mathbf{B}^{\mathrm{T}} \, \mathbf{D} \, \mathbf{B} \, dvol \, \mathbf{d}$$

hence

$$\mathbf{p} = \int \mathbf{B}^{\mathrm{T}} \, \mathbf{D} \, \mathbf{B} \, dvol \, \mathbf{d} \tag{8.38}$$

The expression $\displaystyle\int \mathbf{B}^{\mathrm{T}} \, \mathbf{D} \, \mathbf{B} \, dvol$ is the required element stiffness matrix — $\mathbf{k_n}$.

For the uniform bar element of Fig. 8.6

$$\mathbf{B} = \mathbf{Q}\ \mathbf{A}^{-1} = \begin{bmatrix} 0 & 1 \end{bmatrix} \begin{bmatrix} 1 & 0 \\ -\dfrac{1}{L} & \dfrac{1}{L} \end{bmatrix} = \begin{bmatrix} -\dfrac{1}{L} & \dfrac{1}{L} \end{bmatrix}$$

$$\mathbf{B}^{\mathrm{T}}\ \mathbf{DB} = E \begin{bmatrix} \dfrac{1}{L^2} & -\dfrac{1}{L^2} \\ -\dfrac{1}{L^2} & \dfrac{1}{L^2} \end{bmatrix}$$

therefore

$$\mathbf{k_n} = \int \mathbf{B}^{\mathrm{T}}\ \mathbf{D}\ \mathbf{B}\ \mathrm{d}vol = \mathbf{B}^{\mathrm{T}}\ \mathbf{D}\ \mathbf{B} \int \mathrm{d}vol = \mathbf{B}^{\mathrm{T}}\ \mathbf{D}\ \mathbf{B}\ A\ L$$

$$= \begin{bmatrix} \dfrac{E\ A}{L} & -\dfrac{E\ A}{L} \\ -\dfrac{E\ A}{L} & \dfrac{E\ A}{L} \end{bmatrix}$$

In this case the **B** matrix is constant and **B^{T}DB** is outside the integration. Normally the terms of **B^{T}DB** include variables and need to be integrated for each term in the matrix.

8.5.3   The shape function method
This is the most common approach for type B elements. It is a variant of the method described in section 8.5.2 being a different way of manipulating the basic assumptions.

The basis of the method is that the displacement functions are written directly in terms of the nodal displacements.

If equation (8.33) is substituted into equation (8.31) we get:

$$\mathbf{u} = \mathbf{F}\ \mathbf{A}^{-1}\ \mathbf{d} = \mathbf{N}\ \mathbf{d} \tag{8.39}$$

where **N** is the shape function matrix.

For the uniform bar element of Fig. 8.6 equation (8.39) has the form

$$u = \begin{bmatrix} \left(1 - \dfrac{x}{L}\right) & \left(1 + \dfrac{x}{L}\right) \end{bmatrix} \begin{Bmatrix} u_1 \\ u_2 \end{Bmatrix}$$

i.e

$$u = [N_1\ N_2]\mathbf{d} \tag{8.40}$$

N_1 and N_2 are shape functions. The main property of a shape function is that it has the value 1.0 at the freedom to which it corresponds and the value zero at all other freedoms. Fig. 8.10 shows the shape functions for the uniform bar element.

From equation 8.40 the **B** matrix (defined in equation (8.35)) can be readily established by operating on the shape functions. In this case

$$\varepsilon = \frac{\partial u}{\partial x} = \begin{bmatrix} -\dfrac{1}{L} & \dfrac{1}{L} \end{bmatrix} \mathbf{d} \tag{8.41}$$

This corresponds with the **B** matrix derived from a displacement function in section 8.5.2. The stiffness matrix can now be established using equation (8.38). The advantage of the shape function approach is that the **A** matrix does not need to be inverted. The shape functions are normally written down directly requiring special knowledge for the more complex functions.

8.5.4   Local coordinates and numerical integration

Shape functions are relatively easy to establish for low order elements with rectangular geometry. In order to set up quadrilateral elements it is normal to use local coordinate systems and to carry out a coordinate transformation to the Cartesian set. By doing this one can establish the shape functions as if the element had a rectangular geometry and the transformation maps this to the quadrilateral system. This mapping process uses some mathematical techniques which are normally unfamiliar to structural engineers and one might take this part of the processing for granted. However, there are some aspects which make a general understanding of the process worthwhile.

For illustration, the development of the stiffness matrix for the eight degree of freedom plane stress quadrilateral element shown in Fig. 8.11 is used.

Local coordinates ξ, η are defined which are related to the sides of the element. They have the value 1.0 or -1.0 on the element sides. The origin is where the lines which connect the bisectors of the sides intersect. $\eta = 0.5$ for example is represented by a straight line which connects the $\eta = 0.5$ points on sides 1,4 and 2,3 respectively.

The relationship between the x,y coordinates and the ξ, η coordinates is as follows:

$$x = \tfrac{1}{4}(1 + \xi)(1 + \eta)x_1 + \tfrac{1}{4}(1 - \xi)(1 + \eta)x_2 + \\ \tfrac{1}{4}(1 - \xi)(1 - \eta)x_3 + \tfrac{1}{4}(1 + \xi)(1 - \eta)x_4$$

i.e.

$$x = N_1 x_1 + N_2 x_2 + N_3 x_3 + N_4 x_4$$

similarly

$$y = N_1 y_1 + N_2 y_2 + N_3 y_3 + N_4 y_4 \tag{8.42}$$

where x_i, y_i are the x- and y-coordinates of node i.

Equations (8.42) define the shape of the element in relation to the x-, y-coordinate system.

The displacements are defined using the same functions, i.e

$$u = N_1 u_1 + N_2 u_2 + N_3 u_3 + N_4 u_4$$

and

$$v = N_1 v_1 + N_2 v_2 + N_3 v_3 + N_4 v_4 \tag{8.43}$$

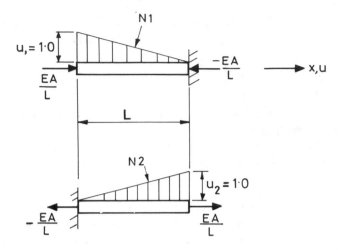

Fig. 8.10 — Shape function for the uniform bar element.

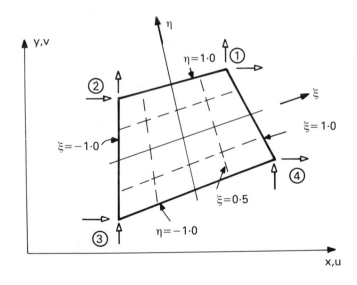

Fig. 8.11 — Local coordinate system for four node plane stress quadrilateral element.

where u_i, v_i are the x and y displacements at node i.

This is an example of an isoparametric element where the same functions are used to define the element shape and the displacements — section 3.5.2.

Equations (8.43) are written in matrix form as

$$\begin{Bmatrix} u \\ v \end{Bmatrix} = \mathbf{N}\,\mathbf{d}$$

where

$$\mathbf{N} = \begin{bmatrix} N_1 & N_2 & N_3 & N_4 & 0 & 0 & 0 & 0 \\ 0 & 0 & 0 & 0 & N_1 & N_2 & N_3 & N_4 \end{bmatrix}$$

and

$$\mathbf{d} = [u_1\ u_2\ u_3\ u_4\ v_1\ v_2\ v_3\ v_4]^T$$

The stiffness matrix is now established in the form:

$$\mathbf{k_e} = \int \mathbf{B^T}\,\mathbf{D}\,\mathbf{B}\ dvol$$

The **B** matrix (which relates ε to $\mathbf{d}$ — see equation (8.35)) is defined by:

$$\begin{Bmatrix} \dfrac{\partial u}{\partial x} \\[2mm] \dfrac{\partial v}{\partial y} \\[2mm] \dfrac{\partial u}{\partial y} + \dfrac{\partial v}{\partial x} \end{Bmatrix} = \begin{bmatrix} \dfrac{\partial N_1}{\partial x} & \dfrac{\partial N_2}{\partial x} & \dfrac{\partial N_3}{\partial x} & \dfrac{\partial N_4}{\partial x} & & & & \\[2mm] 0 & 0 & 0 & 0 & \dfrac{\partial N_1}{\partial y} & \dfrac{\partial N_2}{\partial y} & \dfrac{\partial N_3}{\partial y} & \dfrac{\partial N_4}{\partial y} \\[2mm] \dfrac{\partial N_1}{\partial y} & \dfrac{\partial N_2}{\partial y} & \dfrac{\partial N_3}{\partial y} & \dfrac{\partial N_4}{\partial y} & \dfrac{\partial N_1}{\partial x} & \dfrac{\partial N_2}{\partial x} & \dfrac{\partial N_3}{\partial x} & \dfrac{\partial N_4}{\partial x} \end{bmatrix} \mathbf{d} \qquad (8.44)$$

i.e.

$$\varepsilon = \mathbf{B}\,\mathbf{d}.$$

The problem now is that the N functions are in terms of ξ and η of which x and y are functions. To allow for this change in coordinate system the derivatives of N with respect to ξ and η are written in the form

$$\begin{Bmatrix} \dfrac{\partial \mathbf{N}}{\partial \xi} \\[2mm] \dfrac{\partial \mathbf{N}}{\partial \eta} \end{Bmatrix} = \begin{bmatrix} \dfrac{\partial \mathbf{N}}{\partial x}\dfrac{\partial x}{\partial \xi} + \dfrac{\partial \mathbf{N}}{\partial y}\dfrac{\partial y}{\partial \xi} \\[2mm] \dfrac{\partial \mathbf{N}}{\partial x}\dfrac{\partial x}{\partial \eta} + \dfrac{\partial \mathbf{N}}{\partial y}\dfrac{\partial y}{\partial \eta} \end{bmatrix} = \begin{bmatrix} \dfrac{\partial x}{\partial \xi} & \dfrac{\partial y}{\partial \xi} \\[2mm] \dfrac{\partial x}{\partial \eta} & \dfrac{\partial y}{\partial \eta} \end{bmatrix} \begin{Bmatrix} \dfrac{\partial \mathbf{N}}{\partial x} \\[2mm] \dfrac{\partial \mathbf{N}}{\partial y} \end{Bmatrix}$$

i.e.

$$\begin{Bmatrix} \dfrac{\partial \mathbf{N}}{\partial \xi} \\[2mm] \dfrac{\partial \mathbf{N}}{\partial \eta} \end{Bmatrix} = \mathbf{J} \begin{Bmatrix} \dfrac{\partial \mathbf{N}}{\partial x} \\[2mm] \dfrac{\partial \mathbf{N}}{\partial y} \end{Bmatrix} \qquad (8.45)$$

The **J** matrix in equation (8.45) is called the 'Jacobian'.

The relationship of equation (8.45) is inverted to give:

$$\left\{\begin{array}{c} \dfrac{\partial \mathbf{N}}{\partial x} \\[2mm] \dfrac{\partial \mathbf{N}}{\partial y} \end{array}\right\} = \mathbf{J}^{-1} \left\{\begin{array}{c} \dfrac{\partial \mathbf{N}}{\partial \xi} \\[2mm] \dfrac{\partial \mathbf{N}}{\partial \eta} \end{array}\right\} \tag{8.46}$$

The stiffness matrix related to the ξ, η coordinates is then formed from:

$$\mathbf{k} = \int \mathbf{B}^{\mathrm{T}} \, \mathbf{D} \, \mathbf{B} \, |\mathbf{J}| \, \mathrm{d}\xi \, \mathrm{d}\eta \tag{8.47}$$

where $|\mathbf{J}|$ is the determinant of the Jacobian.

The above integration cannot normally be done explicitly and therefore numerical integration is needed. Normally Gauss quadrature is used. The function to be integrated is calculated at a number of points, multiplied by a weighting factor and the results summed.

For example, in four-point Gauss quadrature, the points used (i.e. the Gauss points) are shown in Fig. 8.12.

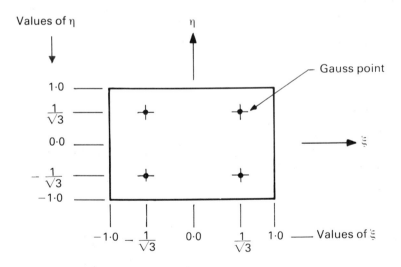

Fig. 8.12 — Gauss points for a rectangular element.

Other types of local coordinate systems are used. For example, area coordinates are used for triangles. These are described in more detail in texts on finite element theory (Zienkiewicz 1977, Dawe 1984, Cook 1981, Ross 1985).

8.6 SOLUTION PROCESS

The basic flow diagram of a stiffness method analysis is shown in Fig. 8.13.

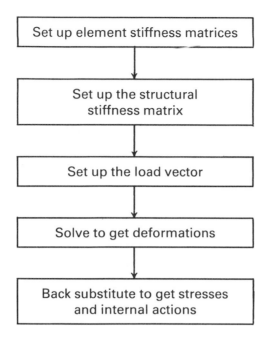

Fig. 8.13 — Flow diagram for the stiffness method.

The text so far has concentrated on elements since that is more important for understanding of modelling techniques. In this section, basic descriptions of the other parts of the process are described.

8.6.1 Setting up the structural stiffness matrix

This part of the process is illustrated using the simple two span continuous beam shown in Fig. 8.14(a). The nodes are numbered (1 to 3) and there are two uniform bending elements a and b — Figs 8.14(b) and (c).

The stiffness relationships for elements a and b are:

$$\left\{ \begin{matrix} m_{a1} \\ m_{a2} \end{matrix} \right\} = \frac{EI}{L} \begin{bmatrix} 4 & 2 \\ 2 & 4 \end{bmatrix} \left\{ \begin{matrix} \theta_{a1} \\ \theta_{a2} \end{matrix} \right\} \text{ i.e. } \mathbf{p_a = k_a d_a}$$

$$\left\{ \begin{matrix} m_{b1} \\ m_{b2} \end{matrix} \right\} = \frac{EL}{L} \begin{bmatrix} 4 & 2 \\ 2 & 4 \end{bmatrix} \left\{ \begin{matrix} \theta_{b1} \\ \theta_{b2} \end{matrix} \right\} \text{ i.e. } \mathbf{p_b = k_b d_b}$$

These are written in diagonally partitioned form

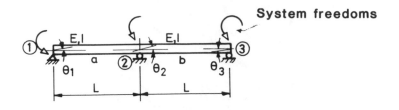

(a) TWO SPAN BEAM

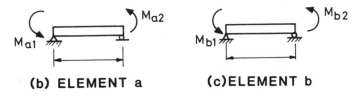

(b) ELEMENT a   **(c)ELEMENT b**

Fig. 8.14 — Continuous beam example.

$$\begin{Bmatrix} m_{a1} \\ m_{a2} \\ m_{b1} \\ m_{b2} \end{Bmatrix} = \frac{EI}{L} \begin{bmatrix} 4 & 2 & 0 & 0 \\ 2 & 4 & 0 & 0 \\ 0 & 0 & 4 & 2 \\ 0 & 0 & 2 & 4 \end{bmatrix} \begin{Bmatrix} \theta_{a1} \\ \theta_{a2} \\ \theta_{b1} \\ \theta_{b2} \end{Bmatrix}$$

i.e.

$$\begin{Bmatrix} \mathbf{m_a} \\ \mathbf{m_b} \end{Bmatrix} = \begin{bmatrix} \mathbf{k_a} & 0 \\ 0 & \mathbf{k_b} \end{bmatrix} \begin{Bmatrix} \mathbf{d_a} \\ \mathbf{d_b} \end{Bmatrix}$$

i.e.

$$\mathbf{p_e} = \mathbf{k_e}\, \mathbf{d_e} \qquad\qquad (8.48)$$

Connection matrix method

The system deformations are $\boldsymbol{\Delta}^\mathbf{T} = \{\theta_1\ \theta_2\ \theta_3\}$

The compatibility condition is

$$\begin{Bmatrix} \theta_{a1} \\ \theta_{a2} \\ \theta_{b1} \\ \theta_{b2} \end{Bmatrix} = \begin{bmatrix} 1 & 0 & 0 \\ 0 & 1 & 0 \\ 0 & 1 & 0 \\ 0 & 0 & 1 \end{bmatrix} \begin{Bmatrix} \theta_1 \\ \theta_2 \\ \theta_3 \end{Bmatrix} \qquad\qquad (8.49)$$

i.e.

$$\mathbf{d_e} = \mathbf{C}\, \boldsymbol{\Delta}$$

Δ is the vector of system deformations
C is a connection matrix.
By contragredience (section 8.4)

$$\mathbf{P}_s = \mathbf{C}^T \mathbf{p}_e \tag{8.50}$$

where $\mathbf{P}_s$ and $\mathbf{p}_e$ are the system and element actions respectively.
   Substituting equations (8.49) and (8.50) into equation (8.48) gives

$$\mathbf{P}_s = \mathbf{C}^T \mathbf{k}_e \mathbf{C} \Delta \tag{8.51}$$

i.e.

$$\mathbf{P}_s = \mathbf{K}\Delta$$

where $\mathbf{K} = \mathbf{C}^T \mathbf{k}_e \mathbf{C}$ is the structural stiffness matrix.
   In this case

$$\mathbf{K} = \frac{EI}{L} \begin{bmatrix} 4 & 2 & 0 \\ 2 & 8 & 2 \\ 2 & 4 & 4 \end{bmatrix} \tag{8.52}$$

Equation (8.51) is a concise way of defining the principle of setting up a structural stiffness matrix demonstrating the principles used, but in computer programs the 'direct stiffness method' is conventionally used.

Direct stiffness method

This is the normal approach because of computational efficiency. The principle is that each element stiffness matrix is taken one at a time and added to the appropriate location in the structural stiffness matrix.

   For the problem of Fig. 8.14 the process would be as shown in Fig. 8.15. The structural stiffness $\mathbf{K}$ has three freedoms and a 3×3 array is created for this purpose. The stiffness matrix for element 'a' is formed and added to the rows and columns 1 and 2 as shown. The stiffness matrix for element 'b' is then formed and added to rows 2 and 3 of $\mathbf{K}$. The same result for $\mathbf{K}$ as given by equation (8.52) is achieved.

   Various algorithms are used for doing this where the element freedoms are treated as nodal groups or individually. For larger problems the structural stiffness and the solution of the equations proceeds in parallel.

   The example of Fig. 8.14 uses elements which have restraints and therefore equation (8.51) can be solved directly. Normally element matrices refer to unres trained coordinates and a structural stiffness formed from them directly would be singular. The structure needs supports which are at least statically determinate to make solutions feasible.

   Basic ways of imposing restraints are:

- Remove the row and column of the structural stiffness matrix which corresponds to a restrained freedom. For example, if Node 3 of the two-span beam of Fig. 8.15 is to be restrained then the structural stiffness matrix is reduced to 2×2 as shown in Fig. 8.16.
- Add a support spring of very high stiffness at the freedom to be restrained. This is

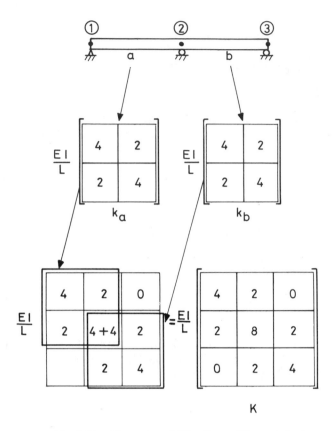

Fig. 8.15 — Formation of K by direct stiffness method.

achieved by increasing the size of the corresponding leading diagonal. The spring stiffness must be such that the movement at the supported freedom has a negligible effect on the stresses in the system. The very stiff support spring does not cause ill conditioning problems (section 5.2.1).

8.6.2 The load vector

Equation (8.51) is the structural stiffness equation to be solved.

Where the loading is applied at the nodes of the model then these loads are simply added to the appropriate locations of $\mathbf{P}_s$. The concept of equivalent joint loads for type A elements is discussed in section 3.4.5. For type B elements loads which are distributed over element, volume, surfaces or edges, i.e. 'body forces', are transformed into a 'consistent' load vector.

The body force vector is denoted as $\mathbf{b}$. We wish to define a vector of nodal forces $\mathbf{p}$ which will be statically equivalent to $\mathbf{b}$. The 'consistent' name means that the tranformation from $\mathbf{b}$ to $\mathbf{p}$ uses the same functions as are used in the formulation of the element stiffness. In terms of shape functions (section 8.5.3) these are, from equation (8.39):

$$\mathbf{u} = \mathbf{N}\,\mathbf{d}$$

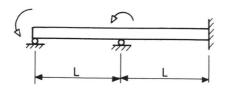

(a) TWO SPAN BEAM

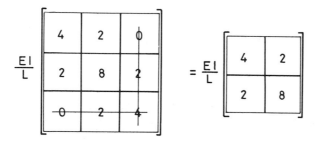

(b) STRUCTURAL STIFFNESS MATRIX

Fig. 8.16 — Continuous beam with fixed end.

where

$\mathbf{u}$ = is the vector of displacement functions
$\mathbf{d}$ = is the vector of nodal displacements

Applying the principle of virtual work to the equilibrium set $\mathbf{b}$ and $\mathbf{p}$ and the corresponding compatible sets $\mathbf{u}$ and $\mathbf{d}$ gives

$$\mathbf{d}^{\mathbf{T}}\,\mathbf{p} = \int \mathbf{u}^{\mathbf{T}}\mathbf{b}\ \mathrm{d}vol$$

Substituting $\mathbf{u}^{\mathbf{T}} = \mathbf{d}^{\mathbf{T}}\,\mathbf{N}^{\mathbf{T}}$ from equation (8.39) gives

$$\mathbf{d}^{\mathbf{T}}\mathbf{p} = \int \mathbf{d}^{\mathbf{T}}\,\mathbf{N}^{\mathbf{T}}\,\mathbf{b}\ \mathrm{d}vol$$

i.e.

$$\mathbf{p} = \int \mathbf{N}^{\mathbf{T}}\,\mathbf{b}\ \mathrm{d}vol \qquad\qquad (8.53)$$

Equation (8.53) provides a rule for establishing the contribution of body forces within an element to the load vector.

8.6.3   The solution routine

The normal approach to solving structural stiffness equations is to use Gaussian elimination. This is the standard approach to solving simultaneous equations where one variable at a time is eliminated to produce a triangular form of the original matrix which can then be solved against the load vector. The symmetry of the structural stiffness equations is taken into account. Another approach is to use the Choleski square root reduction algorithm in which the same general principles are used but a square root is required to be found to establish the leading diagonal term of the reduced matrix.

Structural stiffness matrices are normally 'sparse', i.e. they have a high proportion of terms which are zero and advantage is taken of this normally in one of the following ways.

Band matrices

Depending on how the nodes are numbered, the non zero coefficients of the structural stiffness matrix **K** tend to be close to the leading diagonal of the matrix. The maximum number of columns from the leading diagonal to the last non zero term in any row of the matrix is the half band width of the matrix. This is proportional to the maximum difference between node numbers which are associated with an element. Only the terms within the band are stored, normally making a significant reduction in storage requirements. Also the operations to carry out the solution are confined to the band, thus greatly reducing the solution time.

The matrix of equation (8.52) is a band matrix. Fig. 8.17 shows how it can be represented in half band form. The upper band is shown in Fig. 8.17(b). The lower part of the band need not be stored because **K** is symmetric. There would not be a significant storage or computational advantage in this case but in large problems the improvement in efficiency can be very significant. Automatic renumbering of nodes to optimise the bandwidth is a common feature of analysis packages.

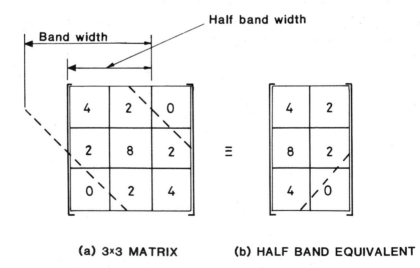

(a) 3×3 MATRIX   **(b) HALF BAND EQUIVALENT**

Fig. 8.17 — Example of band matrix.

Frontal solution

For frontal solutions (sometimes called 'wavefront' or 'skyline' solutions) the equations are processed in an order that is affected by the element numbering rather than the node numbering. The 'frontwidth' is like a localized bandwidth which changes as the processing proceeds.

The basis of the solution remains as Gaussian elimination and the term 'frontal solution' refers to the algorithms used to reduce storage and computation time. Automatic renumbering of the elements is a common feature for packages which use this technique.

8.6.4   Back substitution

To obtain internal forces or element stresses a back substitution is made in the constitutive relationship, i.e. using equations (2.2) and (8.35):

$$\boldsymbol{\sigma} = \mathbf{D}\,\boldsymbol{\varepsilon} = \mathbf{D}\,\mathbf{B}\,\mathbf{d_e}$$

where $\mathbf{d_e}$ is the vector of element nodal deformations extracted from the solution vector $\boldsymbol{\Delta}$. The back substitution matrix $\mathbf{DB}$ may be either stored when the element stiffness is being formulated or reformed at the back substitution stage.

8.7   DERIVATIONS OF ELASTIC CONSTITUTIVE RELATIONSHIPS

8.7.1   Plane stress

Direct stress terms

The conventional elastic constants are based on measurements made in a tensile test, i.e under plane stress conditions. Fig. 8.18 shows x and y-specimens cut from a plate.

An axial load P is applied to the x-specimen resulting in a direct stress $\sigma_x = P/A$. The strains ε_x, ε_y in the x- and y-directions are measured and Young's modulus and Poisson's ratio are established on the basis of:

$$\varepsilon_x = \sigma_x/E_x, \qquad \varepsilon_y = -\nu_{xy}\varepsilon_x \tag{8.54}$$

Similar measurements on the y-specimen will give

$$\varepsilon_y = \sigma_y/E_y, \qquad \varepsilon_x = -\nu_{yx}\varepsilon_y \tag{8.55}$$

If the material is isotropic (i.e. if it has the same properties in all directions) then:

$$E_x = E_y = E \text{ and } \nu_{xy} = \nu_{yx} = \nu \tag{8.56}$$

Substituting equation (8.56) into equations (8.54) and (8.55) and writing in matrix notation gives:

$$\begin{Bmatrix} \varepsilon_x \\ \varepsilon_y \end{Bmatrix} = \frac{1}{E} \begin{bmatrix} 1 & -\nu \\ -\nu & 1 \end{bmatrix} \begin{Bmatrix} \sigma_x \\ \sigma_y \end{Bmatrix} \tag{8.57}$$

Inverting equation (8.57) gives

$$\begin{Bmatrix} \sigma_x \\ \sigma_y \end{Bmatrix} = \frac{E}{1-\nu^2} \begin{bmatrix} 1 & \nu \\ \nu & 1 \end{bmatrix} \begin{Bmatrix} \varepsilon_x \\ \varepsilon_y \end{Bmatrix} \tag{8.58}$$

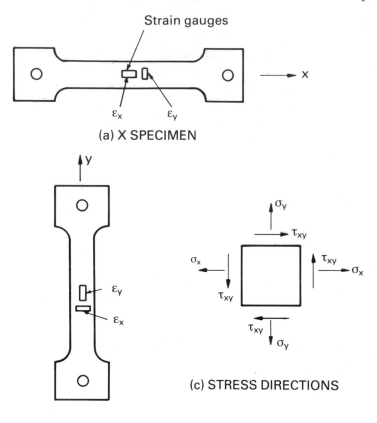

Fig. 8.18 — Axial tension specimens.

Rigidity modulus — G

Fig. 8.19(a) shows an element of linear elastic material with side lengths d. It is subject to equal and opposite stresses σ acting in the x- and y-directions. Fig. 8.19(b) is the Mohr circle from this situation. On planes at 45° to the x- and y-axes the stress state is 'pure shear' with shear stress τ and no direct stress on these planes.

Fig. 8.19(c) shows the deformation of the element relative to the pure shear planes.

θ is the angle through which one of the sides moves and therefore

$$\gamma = 2\theta \tag{8.59}$$

where γ is the shear strain

The element moves a distance 'a' outwards on either side of the axis on symmetry in the x-direction and correspondingly moves a distance of 'a' inwards in the y direction. Therefore

$$\varepsilon_x = \frac{2a}{d} \tag{8.60}$$

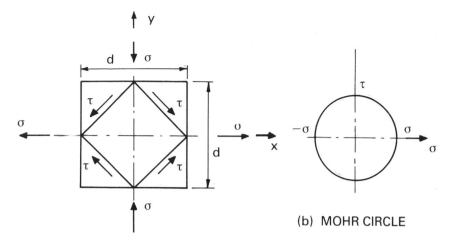

(a) BIAXIAL STRESS
System to give pure
shear on 45° planes

(b) MOHR CIRCLE

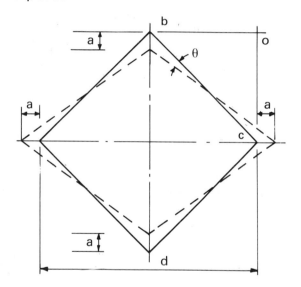

(c) DEFORMED SHAPE OF 45° ELEMENT

Fig. 8.19 — Element in pure shear.

Three conditions need to be satisfied
● Force deformation, in the x-direction

$$\varepsilon_x = \frac{1}{E}\left(\sigma - \nu(-\sigma)\right) = \frac{\sigma}{E}\left(1 + \nu\right) \qquad (8.61)$$

- **Equilibrium, from the Mohr circle**

$$\tau = \sigma \tag{8.62}$$

- **Compatibility, the point 'o' on Fig. 8.19(b) is the centre of rotation of side** *bc*

Therefore

$$\theta = \frac{a}{d/2} = \frac{2a}{d}$$

Therefore

$$\gamma = 2\theta = \frac{4a}{d} \tag{8.63}$$

Hence from equation (8.60)

$$\varepsilon_x = \frac{\gamma}{2} \tag{8.64}$$

Substituting equations (8.62) and (8.64) into equation (8.61) therefore gives

$$\tau = \frac{E}{2(1+v)}\, \gamma$$

hence

$$G = \frac{E}{2(1+v)} \tag{8.65}$$

Plane stress matrix
Putting equations (8.58) and (8.65) together gives the isotropic plane stress relationship as given — equation (2.3).

8.7.2　Plane strain
For plane strain the condition $\varepsilon_z = 0$ is imposed on the plane stress situation.

The direct stress and strain in the z-direction are added to equation (2.6) giving:

$$\begin{Bmatrix} \varepsilon_x \\ \varepsilon_y \\ \varepsilon_z \end{Bmatrix} = \frac{1}{E} \begin{bmatrix} 1 & -v & -v \\ -v & 1 & -v \\ -v & -v & 1 \end{bmatrix} \begin{Bmatrix} \sigma_x \\ \sigma_y \\ \sigma_z \end{Bmatrix} \tag{8.66}$$

If $\varepsilon_z = 0$ then from the last line of equation (8.66) one gets

$$\sigma_z = v(\sigma_x + \sigma_y) \tag{8.67}$$

Substituting this expression into the first two lines of equation (8.66), inverting the resulting 2×2 matrix and adding the shear term gives the plane strain constitutive relationship, equation (2.4).

8.7.3   Orthotropic plane stress

Orthotropic plane stress conditions are developed from equations (8.54) and (8.55). The direct stress–strain relationship is:

$$
\left\{ \begin{array}{c} \varepsilon_x \\ \\ \varepsilon_y \end{array} \right\} = \begin{bmatrix} \dfrac{1}{E_x} & -\dfrac{v_{xy}}{E_x} \\ \\ -\dfrac{v_{yx}}{E_y} & \dfrac{1}{E_y} \end{bmatrix} \left\{ \begin{array}{c} \sigma_x \\ \\ \sigma_y \end{array} \right\}
\tag{8.68}
$$

This is inverted to give:

$$
\left\{ \begin{array}{c} \sigma_x \\ \sigma_y \end{array} \right\} = \frac{1}{(1 - v_{xy}v_{yx})} \begin{bmatrix} E_x & v_{xy}E_y \\ v_{yx}E_x & E_y \end{bmatrix} \left\{ \begin{array}{c} \varepsilon_x \\ \varepsilon_y \end{array} \right\}
\tag{8.69}
$$

The orthotropic plane stress constitutive relationship is therefore: (8.70)

$$
\left\{ \begin{array}{c} \sigma_x \\ \sigma_y \\ \tau_{xy} \end{array} \right\} = \begin{bmatrix} E'_x & E'' & 0 \\ E'' & E'_y & 0 \\ 0 & 0 & G \end{bmatrix} \left\{ \begin{array}{c} \varepsilon_x \\ \varepsilon_y \\ \gamma_{xy} \end{array} \right\}
\tag{8.70}
$$

where

$$E'_x = E_x/(1 - v_{xy}v_{yx})$$

$$E'_y = E_y/(1 - v_{xy}v_{yx})$$

$$E'' = \frac{v_{xy}E_y}{(1 - v_{xy}v_{yx})} = \frac{v_{yx}E_x}{(1 - v_{xy}v_{yx})}$$

8.7.4   Isotropic 3D

The direct stress-strain part of this relationship is obtained by inverting equation (8.66) and the shear equations have only diagonal terms. The resulting matrix is given as equation (2.6).

8.7.5   Thin plate bending

In this section the constitutive relationship for this plate bending is derived. The sign conventions are those used in Timoshenko & Woinowsky-Krieger (1959) — see Fig. 2.4.

Direct moments

Fig. 8.20(a) shows a thin plate under biaxial bending and Fig. 8.20(b) shows section AA through the plate. The normal to the plate remains straight but rotates through an angle $\partial w/\partial x$ (the normal being at right angles to the tangent to the displacement curve).

From the geometry of Fig. 8.20(b),

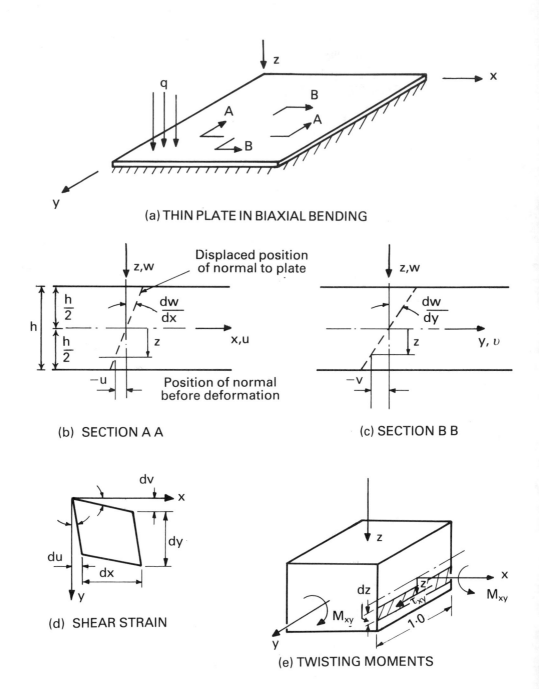

(a) THIN PLATE IN BIAXIAL BENDING

(b) SECTION A A

(c) SECTION B B

(d) SHEAR STRAIN

(e) TWISTING MOMENTS

Fig. 8.20 — Thin plate bending.

$$-\frac{u}{z} = \frac{\partial w}{\partial x}$$

therefore

$$u = -z\frac{\partial w}{\partial x} \qquad (8.71)$$

therefore

$$\varepsilon_x = \frac{\partial u}{\partial x} = -z\frac{\partial^2 w}{\partial x^2} \qquad (8.72)$$

Similarly from Fig. 8.20(c)

$$\upsilon = -z\frac{\partial w}{\partial y} \qquad (8.73)$$

therefore

$$\varepsilon_y = \frac{\partial \upsilon}{\partial y} = -z\frac{\partial^2 w}{\partial x^2} \qquad (8.74)$$

i.e.

$$\left\{\begin{array}{c} \varepsilon_x \\ \\ \varepsilon_y \end{array}\right\} = \left\{\begin{array}{c} -z\dfrac{\partial^2 w}{\partial x^2} \\ \\ -z\dfrac{\partial^2 w}{\partial y^2} \end{array}\right\} \qquad (8.75)$$

The bending moment per unit width on the x-plane is:

$$M_x = \int_{-h/2}^{h/2} \sigma_x \, z \, dz$$

where h is the plane thickness and on the y-plane

$$M_y = \int_{-h/2}^{h/2} \sigma_y \, z \, dz$$

Therefore

$$\left\{\begin{array}{c} M_x \\ M_y \end{array}\right\} = \int_{-h/2}^{h/2} \left[\begin{array}{c} \sigma_x z\,dz \\ \sigma_y z\,dz \end{array}\right] \qquad (8.76)$$

Substituting equation (8.75) into equation (8.58), substituting the result of this into equation (8.76) and performing the integration gives:

$$\left\{ \begin{array}{c} M_x \\ \\ M_y \end{array} \right\} = D \left[\begin{array}{cc} 1 & v \\ & \\ v & 1 \end{array} \right] \left\{ \begin{array}{c} -\dfrac{\partial^2 w}{\partial x^2} \\ \\ -\dfrac{\partial^2 w}{\partial y^2} \end{array} \right\} \tag{8.77}$$

where

$$D = \frac{Eh^3}{12(1 - v^2)} \tag{8.78}$$

Twisting moments

Fig. 8.20(d) shows the shear deformation of a differential element of the plate. Shear strain

$$\gamma_{xy} = \frac{\partial u}{\partial y} + \frac{\partial v}{\partial x}$$

Substituting from equations (8.71) and (8.73) gives

$$\gamma_{xy} = -2 z \frac{\partial^2 w}{\partial x \partial y} \tag{8.79}$$

Shear stress $\tau_{xy} = G\, \gamma_{xy} = -2G\, z\dfrac{\partial^2 w}{\partial x \partial y}$ \hfill (8.80)

The twisting moment is defined as (Fig. 8.20(e)):

$$M_{xy} = -\int_{-h/2}^{h/2} \tau_{xy}\, z\, \mathrm{d}z \tag{8.81}$$

M_{xy} is positive anticlockwise on the x-face. M_{yx} is then positive clockwise on the y-face for equilibrium of the element. The sign of the twisting moment is defined by its action on the x-plane For positive z, τ_{xy} is in the y-direction, producing a clockwise moment on the x-face. Hence the negative sign in equation (8.81).

   Substituting equation (8.80) into equation (8.81), substituting for G (equation (8.65)) and performing the integration gives:

$$M_{xy} = D(1 - v) \frac{\partial^2 w}{\partial x y} \tag{8.82}$$

Thin plate bending matrix

The complete constitutive matrix (quoted as equation (2.8)) is therefore:

$$\left\{ \begin{matrix} M_x \\ M_y \\ M_{xy} \end{matrix} \right\} = D \begin{bmatrix} 1 & v & 0 \\ v & 1 & 0 \\ 0 & 0 & \dfrac{1+v}{2} \end{bmatrix} \left\{ \begin{matrix} -\dfrac{\partial^2 w}{\partial x^2} \\ -\dfrac{\partial^2 w}{\partial y^2} \\ 2\dfrac{\partial^2 w}{\partial x^2} \end{matrix} \right\} \tag{8.83}$$

8.8   DERIVATION OF GEOMETRIC STIFFNESS MATRICES

8.8.1   Beam elements

Fig. 8.21 shows a differential element of a beam with lateral displacement dv and axial load N. The moment due to the eccentricity is Ndv and the corresponding rotation is dv/dx. Expressing this moment as $N(dv/dx)dx$ gives the internal virtual work on the differential element as:

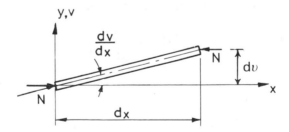

Fig. 8.21 — Differential element of beam with eccentric axial load.

$$\text{Internal work} = \frac{dv}{dx} N \frac{dv}{dx} \, dx = N\left(\frac{dv}{dx}\right)^2 dx \tag{8.84}$$

The lateral displacement is written in the same form as equation (8.31) i.e.

$$v = \mathbf{F}\boldsymbol{\alpha} \tag{8.85}$$

where equation (8.85) is a matrix form of equation (3.4).
Therefore

$$\frac{dv}{dx} = \mathbf{G}\alpha \tag{8.86}$$

where

$$\mathbf{G} = \frac{d}{dx}\,\mathbf{F}$$

Substituting in equation (8.86) for α from equation (8.33) gives

$$\frac{dv}{dx} = \mathbf{G}\,\mathbf{A}^{-1}\mathbf{d}$$

$$= \mathbf{Ed}$$

where

$$\mathbf{E} = \mathbf{GA}^{-1} \tag{8.87}$$

Equation (8.87) is now substituted in equation (8.84). Since the internal work is a scalar quantity the scalar product for $(dv/dx)^2$ must be used, i.e.

$$\text{Internal work} = \left\{\frac{dv}{dx}\right\}^{T} N\left\{\frac{dv}{dx}\right\}\, dx \tag{8.88}$$

The virtual work equation corresponding to equation (8.37) is therefore

$$\mathbf{d}^{T}\mathbf{p} = \int \varepsilon^{T}\sigma\ dvol + \int\ \left\{\frac{dv}{dx}\right\}^{T} N\left\{\frac{dv}{dx}\right\}\ dx \tag{8.89}$$

Substituting equation (8.35) and the constitutive relationship equation 2.7 into equation (8.89) and cancelling $\mathbf{d}^{T}$ gives

$$\mathbf{p} = [\int \mathbf{B}^{T}\ \mathbf{D}\ \mathbf{B}\ dx + \int \mathbf{E}^{T}\ N\ \mathbf{E} dx]\mathbf{d} \tag{8.90}$$

The first term on the right-hand side of equation (8.90) is the elastic bending stiffness matrix and the second term represents a 'geometric' stiffness matrix which takes account of the eccentricity of the applied load — see section 3.11.

The geometric stiffness matrix reduces the element stiffness and causes equation (8.90) to be non-linear requiring an iterative solution as discussed in section 4.8.

8.8.2 Plate bending elements
In two dimensions the internal work on an element $dxdy$ corresponding to equation (8.84) is (Allen & Bulson 1980).
Internal work =

$$h\left(\sigma_x\left(\frac{\partial w}{\partial x}\right)^2 + \sigma_y\left(\frac{\partial w}{\partial y}\right)^2 + 2\ \tau_{xy}\frac{\partial^2 w}{\partial w\partial y}\right]\ dxdy$$

where h is the element thickness.

This expression can be written in the form

$$\text{Internal work} = \left\{ \begin{matrix} \dfrac{\partial w}{\partial x} \\[2ex] \dfrac{\partial w}{\partial y} \end{matrix} \right\}^{T} \begin{bmatrix} \sigma_x & \tau_{xy} \\[1ex] \tau_{xy} & \chi_y \end{bmatrix} \left\{ \begin{matrix} \dfrac{\partial w}{\partial x} \\[2ex] \dfrac{\partial w}{\partial y} \end{matrix} \right\} \, dxdy$$

i.e. Internal work $= \mathbf{I}^T \, \boldsymbol{\sigma} \, \mathbf{I} \, dxdy$

$\mathbf{I}$ is set up by operating on the equivalent of equation (8.34) and substituting for the equivalent of equation (8.33) giving:

$$\mathbf{I} = \left\{ \begin{matrix} \dfrac{\partial}{\partial x} \\[2ex] \dfrac{\partial}{\partial y} \end{matrix} \right\} w = \left\{ \begin{matrix} \dfrac{\partial}{\partial x} \\[2ex] \dfrac{\partial}{\partial y} \end{matrix} \right\} \mathbf{F} \mathbf{A}^{-1} \mathbf{d} = \mathbf{E} \mathbf{d} \tag{8.91}$$

where $\mathbf{E} = \left\{ \begin{matrix} \dfrac{\partial}{\partial x} \\[2ex] \dfrac{\partial}{\partial y} \end{matrix} \right\} \mathbf{F} \, \mathbf{A}^{-1}$

The virtual work equation corresponding to equation (8.89) is:

$$\mathbf{d}^T \mathbf{p} = \int \boldsymbol{\varepsilon}^T \boldsymbol{\sigma} \, dvol + \int \mathbf{I}^T \, \boldsymbol{\sigma} \, \mathbf{I} \, dxdy \tag{8.92}$$

This leads to the element stiffness expression

$$\mathbf{p} = \left[\int \mathbf{B}^T \mathbf{D} \mathbf{B} \, dvol + \int \mathbf{E}^T \boldsymbol{\sigma} \mathbf{E} \, dxdy \right] \mathbf{d} \tag{8.93}$$

The second term of the right-hand side of equation (8.93) is the plate bending geometric stiffness matrix.

8.9   AN ALGORITHM TO SOLVE THE PLATE ON SPRINGS MODEL

The main thrust of this text is towards use of standard analysis software backed up by 'back of an envelope' checks. This is based on the fact that to get the best out of modern techniques, the programs need man years of effort to write. However, the solution of the plate on springs model described in sections 4.7.3 and 6.6.1 involves a maximum of three unknowns and the process of writing a program to do this may be a useful exercise. The procedure involved has the essential elements of a stiffness analysis. Study of this can promote understanding of the basic principles used in standard analysis software.

In section 6.6.1 the basis of a solution for models with springs parallel to the coordinate axes is described. Here the relationships are developed on the basis of a spring that can be at an angle to the coordinate axes.

Fig. 8.22(a) shows a rigid plate supported on springs. A master node-m-is chosen at a point on the plate.

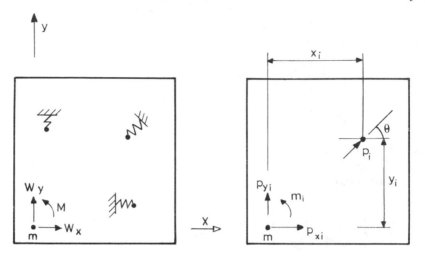

(a) RIGID PLATE   (b) SPRING FORCES

Fig. 8.22 — Rigid plate on spring supports.

Each spring is defined by the relationship

$$p_i = k_i \, \delta_i \tag{8.94}$$

where p_i is the force in the line of the spring, δ_i is the corresponding deformation.

Fig. 8.22(b) shows a spring force p_i and the statically equivalent actions p_{xi}, p_{yi}, m_i at the master node. These actions are related as follows:

$$\left\{ \begin{array}{c} p_{xi} \\ p_{yi} \\ m_i \end{array} \right\} = \left[\begin{array}{c} \cos \theta \\ \sin \theta \\ x_i \sin \theta - y_i \cos \theta \end{array} \right] p_i$$

i.e.

$$\mathbf{p}_{sp} = \mathbf{H}^{\mathrm{T}} \, p_i \tag{8.95}$$

where the subscript 'sp' refers to spring (i.e. element) quantities.

By contragredience (section 8.4) the corresponding deformations are related by:

$$\delta_i = \mathbf{H} \, \mathbf{d}_{sp} \tag{8.96}$$

where $\mathbf{d}_{sp}$ is the spring deformations corresponding to $\mathbf{p}_{sp}$. Substituting equations (8.95) and (8.96) into equation (8.94) gives:

$$\mathbf{p_{sp}} = \mathbf{H^T}\, k_i\, \mathbf{H}\, \mathbf{d_{sp}}$$

i.e.

$$\mathbf{p_{sp}} = \mathbf{k_{sp}}\, \mathbf{d_{sp}} \tag{8.97}$$

where $\mathbf{k_{sp}} = \mathbf{H^T}\, k_i\, \mathbf{H}$ is the spring stiffness related to freedoms at the master node. Note that:

$$\mathbf{k_{sp}} = \begin{bmatrix} \cos^2\theta & -\cos\theta\sin\theta & (x_i\sin\theta\cos\theta - y_i\cos^2\theta) \\ & \sin^2\theta & (x_i\sin^2\theta - y_i\sin\theta\cos\theta) \\ \text{symmetric} & & (x_i\sin\theta - y_i\cos\theta)^2 \end{bmatrix}$$

We now want to establish the system stiffness relationship

$$\mathbf{P_{rp}} = \mathbf{K_{rp}}\, \mathbf{\Delta_{rp}} \tag{8.98}$$

where the subscript 'rp' refers to rigid plate.
    The equilibrium relationship is

$$\mathbf{P_{rp}} = \Sigma\, \mathbf{p_{sp}} \tag{8.99}$$

and compatibility is

$$\mathbf{d_{sp}} = \mathbf{\Delta_{rp}} \tag{8.100}$$

Substituting equation (8.99) and (8.100) into equation (8.97) gives:

$$\mathbf{P_{rp}} = \Sigma\, \mathbf{k_{sp}}\, \mathbf{\Delta_{rp}} \tag{8.101}$$

Thus the system stiffness is $\mathbf{K_{rp}} = \Sigma\mathbf{k_{sp}}$
    The procedure is therefore as follows:

(1) Establish a 3×3 matrix for $\mathbf{K_{rp}}$.
(2) Establish $\mathbf{k_{sp}}$ for a spring and add it to $\mathbf{K_{rp}}$. Repeat for all springs.
(3) Set up the load vector and solve for $\mathbf{\Delta_{rp}}$.
(4) Calculate the spring loads using

$$p_i = k_i\, \mathbf{H}\, \mathbf{\Delta_{rp}} \tag{8.102}$$

Appendix 1: Tables

Table A1 — Second moments of area

	Area	Second moment of area
	BD	$\dfrac{1}{12}BD^3$
	$BD - bd$	$\dfrac{1}{12}[BD^3 - bd^3]$
	$\dfrac{1}{2}bd$	$\dfrac{1}{36}bd^3$
	πR^2	$\dfrac{\pi}{4}R^4$
	$\pi(R^2 - r^3)$	$\dfrac{\pi}{4}(R^4 - r^4)$
	$2\pi Rt$	$\pi R^3 t$

Table A2 — Shear areas for beams

Shape	$\overline{A}$
Reactangle	$5/6\,A$
Solid circle	$1.1\,A$
Thin-walled tube	$0.5\,A$
I-section	$\simeq 1.0\,A_{web}$
T-section	$\simeq 0.85\,A_{web}$

A = cross-sectional area.
$\overline{A}$ = shear area.
A_{web} = area of web

Table A3 — St Venant torsion constant — C

Shape	C
Circular section	C = second polar moment of area = $\pi R^4/2$
Closed thin-walled section	$C = \dfrac{4A_0^2}{\displaystyle\int \dfrac{d_t}{t}}$ A_0 is the area enclosed by a line through the mid-thickness. The integral is carried out along the perimeter.
Narrow rectangle $\dfrac{b}{t} > 5$	b over a rectangle of thickness t $C = \dfrac{1}{3}bt^3$
Open section composed of narrow rectangles	$C = \sum \dfrac{1}{3}bt^3$ b is the larger dimension of a rectangle t is the width of the rectangle The summation is over all rectangles
Closed rectangular thick-walled section	See Marshall (1970)
Solid rectangle	rectangle of height t, width b $C = bt^3\left[\dfrac{1}{3} - 0.21\dfrac{t}{b}\left(1 - \dfrac{t^4}{12b^4}\right)\right]$
Solid square	$b \times b$ square $C = 0.1406b^4$

Table A4 — Typical values for modules of subgrade reaction k_s

Soil	k_s, kN/m^3
Loose sand	4800– 16000
Medium dense sand	9600– 80000
Dense sand	64000–128000
Clayey medium dense sand	32000– 80000
Silty medium dense sand	24000– 48000
Clayey soil:	
$\qquad q_u \leqslant 200\,\text{N/mm}^2$	12000–24000
$200 < q_u \leqslant 400\,\text{N/mm}^2$	24000–48000
$\qquad q_u > 800\,\text{N/mm}^2$	> 48000

q_u = bearing capacity.
From Bowles (1982).

Table A5 — Typical values for modulus of elasticity for soils E_s

Soil	E_s N/mm^2
Clay	
$\qquad$ Very soft	2–15
$\qquad$ Soft	5–25
$\qquad$ Medium	15–50
$\qquad$ Hard	50–100
$\qquad$ Sandy	25–250
Glacial till	
$\qquad$ Loose	10–153
$\qquad$ Dense	144–720
$\qquad$ Very dense	478–1440
Loess	14–57
Sand	
$\qquad$ Silty	7–21
$\qquad$ Loose	10–24
$\qquad$ Dense	48–81
Sand and gravel	
$\qquad$ Loose	48–148
$\qquad$ Dense	96–192
Shale	144–14400
Silt	2–20

From Bowles (1982).

Table A6 — Typical values for Poisson's ratio for soils — υ

Type of soil	υ
Clay, saturated	0.4–0.5
Clay, unsaturated	0.1–0.3
Sandy clay	0.2–0.3
Silt	0.3–0.35
Sand (dense)	0.2–0.4
Coarse (void ratio = 0.4–0.7)	0.15
Fine grained (void ratio = 0.4–0.7)	0.25
Rock	0.1–0.4 (depends on type of rock)
Loess	0.1–0.3
Ice	0.36
Concrete	0.15

From Bowles (1982).

Table A7 — Recommended damping values

Stress level	Type and condition of structure	Percentage critical damping
Working stress, no more than about 1/2 yield point	Welded steel, prestressed concrete, well-reinforced concrete (only slight cracking)	2 to 3
	Reinforced concrete with considerable cracking	3 to 5
	Bolted and/or riveted steel, wood structures with nailed or bolted joints	5 to 7
At or just below yield point	Welded steel, prestressed concrete (without complete loss of prestressing)	5 to 7
	Prestressed concrete with no prestress left	7 to 10
Reinforced concrete	Reinforced concrete	7 to 10
	Bolted and/or riveted steel, wood structures with bolted joints	10 to 15
	Wood structures with nailed joints	15 to 20

From *Earthquake Spectra and Design* — Earthquake Engineering Research Institute.

Table A8 — Deflection formulae for beams

System	Bending moment	Maximum bending deflection	Maximum shear deflection
Cantilever, point load W at free end, span L	WL	$\dfrac{WL^3}{3EI}$	$\dfrac{WL}{\overline{A}G}$
Cantilever, UDL W	$\dfrac{WL}{2}$	$\dfrac{WL^3}{8EI}$	$\dfrac{WL}{2\overline{A}G}$
Simply supported, central point load W	$\dfrac{WL}{4}$	$\dfrac{WL^3}{48EI}$	$\dfrac{WL}{4\overline{A}G}$
Simply supported, UDL W	$\dfrac{WL}{8}$	$\dfrac{5}{384}\dfrac{WL^3}{EI}$	$\dfrac{WL}{8\overline{A}G}$
Fixed both ends, central point load W	$\dfrac{WL}{8}$, $\dfrac{WL}{8}$, $\dfrac{WL}{8}$	$\dfrac{WL^3}{192EI}$	$\dfrac{WL}{4\overline{A}G}$
Fixed both ends, UDL W	$\dfrac{WL}{12}$, $\dfrac{WL}{12}$, $\dfrac{WL}{24}$	$\dfrac{WL^3}{384EI}$	$\dfrac{WL}{8\overline{A}G}$
Propped cantilever, central point load W	$\dfrac{3}{16}WL$, $\dfrac{5}{32}WL$	$\dfrac{WL^3}{107.3EI}$	$\dfrac{WL}{4\overline{A}G}$
Propped cantilever, UDL W	$\dfrac{WL}{8}$, $\dfrac{9}{128}WL$	$\dfrac{WL^3}{185EI}$	$\dfrac{WL}{8\overline{A}G}$
Simply supported with end moments M_A, M_B, UDL W	M_A, $\dfrac{WL}{8}$, M_B	$\dfrac{5}{384}\dfrac{WL^3}{EI}-\dfrac{(M_A+M_B)^2}{16EI}$	$\dfrac{WL}{8\overline{A}G}$
Cantilever, triangular load (max at fixed end) W	$\dfrac{WL}{6}$	$\dfrac{WL^3}{15EI}$	$\dfrac{WL}{3\overline{A}G}$
Cantilever, triangular load (max at free end) W	$\dfrac{WL}{3}$	$\dfrac{11}{16}\dfrac{WL^3}{EI}$	$\dfrac{2}{3}\dfrac{WL}{\overline{A}G}$

Table A9 — Typical physical properties of structural materials

Material	Density (kg/m^3)	Young's modulus E (kN/mm^2)	Poisson's ratio (v)	Coefficient of linear expansion/°C ($\times 10^{-6}$)
Steel	7850	200	0.3	11.5
Cast iron	7100–7500	76–145	0.3	11.0–14.0
Wrought iron	7200	169–176	0.3	10.0–12.5
Aluminium	2700	68–72	0.35	24.0
Concrete				
Normal weight	2360	20–40	0.2	10.0–14.0
Light weight	320–1920	7–19	0.2	3.6–8.0
Timber				
Softwood	500–700	5–9	0.2	4.5
Hardwood	600–1100	8–18		4.5
Brick masonry	2220	6–18		} 5.6 (hori.) 8.0 (vert.)
Stone masonry	2100–3000			3.0–12.0

Appendix 2:
Calculation of shear deformation in beams

Since shear deformation in beams is seldom considered to be important the theory behind its use is not normally included in texts. Shear deformation is important in some beam situations and the concept is useful in checking models as discusssed in Chapter 6. The basic theory is therefore outlined here.

SHEAR DEFORMATION OF A DIFFERENTIAL ELEMENT

Fig. A1(a) shows a cantilever beam with tip load P and retangular cross-section. A differential element (Fig. A1(b)) deforms as in Fig. A1(c) under a shear s_0 ($= P$ in this case).

The shear stress at distance y from the centroidal axis of the section (Fig. A1(d)) is given by:

$$\tau = \frac{s_0 Q}{Ib} \tag{A1}$$

where

$$Q = \left(\frac{d}{2} - y\right)\left(\frac{d}{2} + y\right)\frac{b}{2} \text{ and } I = \frac{bd^3}{12} \text{ in this case.}$$

Therefore

$$\tau = \frac{6s_0}{bd^3}\left(\frac{d^2}{4} - y^2\right) \tag{A2}$$

The shear stress and the shear strain are related by

$$\tau = G\gamma \tag{A3}$$

For the differential element we have an equilibrium set s_0, τ and a corresponding compatible set $d\upsilon$, γ. The principle of virtual work can therefore be applied, i.e.

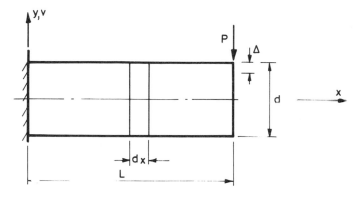

(a) CANTILEVER WITH TIP LOAD

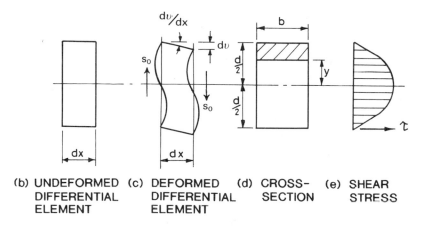

(b) UNDEFORMED   (c) DEFORMED   (d) CROSS-   (e) SHEAR
    DIFFERENTIAL   DIFFERENTIAL   SECTION   STRESS
    ELEMENT   ELEMENT

Fig. A1 — Shear deformation of a beam.

$$s_0 \, dv = \int \tau \gamma \, dx \, dy \, dz \tag{A4}$$

Substituting equation (A3) into equation (A4) and integrating over the width (in the z-direction), the virtual work expression for the differential element is:

$$s_0 \, dv = b \, dx \int \frac{\tau^2}{G} \, dy \tag{A5}$$

Substituting equation (A2) into equation (A5) gives

$$s_0 dv = \frac{36 s_0^2}{b d^6} \, dx \int_{-d/2}^{d/2} \left(\frac{d^2}{4} - y^2 \right)^2 \, dy$$

Performing the integration gives:

$$dv = \frac{s_0}{\frac{5}{6} b \, dG} \, dx = \frac{s_0}{\bar{A} \, G} \, dx \qquad\qquad (A6)$$

where $\bar{A}$ is the shear area $= \frac{5}{6} A$ for a rectangle.

For an I section the shear stess is assumed to be constant within the depth with a value of

$$\tau = \frac{s_0}{A_{\text{web}}}$$

Applying the process used for the rectangle gives:

$$\bar{A} = A_{\text{web}}$$

SHEAR DEFORMATION OF A BEAM

To calculate the shear deformation of a beam, the principle of virtual work can again be used.

The cantilever of Fig. A1(c) has internal shear displacement dv and external displacement at tip Δ. Using a unit virtual tip load for the equilibrium set, the virtual work expression is:

$$1 \cdot \Delta = \int s_1 \, dv$$

where s_1 is the shear force due to the unit load.

Substituting from equation (A6) for dv gives:

$$\Delta = \int s_1 \frac{s_0 \, dx}{\bar{A} \, G} \qquad\qquad (A8)$$

This is the general virtual work expression for shear deformation in beams. It is analogous to the conventional virtual work expression beam bending, axial deformation, etc.

For the cantilever the tip deflection is:

$$\Delta = \int_0^L 1.0 \times \frac{P}{\bar{A} \, G} \, dx = \frac{PL}{\bar{A} \, G}$$

References and further reading

Adini, A. & Clough, R. W. (1961) Analysis of plate bending by the finite element method. *Report to National Science Foundation, USA, No G7337.*

Alexander, S. J. & Hambly, E. C. (1970) The design of structures to withstand gaseous explosions. *Concrete,* **4**, Part 1, Feb., 62–65, Part 2, Mar., 107–116.

Allen, H. G. & Bulson, P. S. (1980) *Background to Buckling.* McGraw-Hill.

Argyris, J. H. (1954) Energy theorems and structural analysis. *Aircraft Engineering,* Vol. 26, pp. 137–383.

ASCE (1985) Design of structures to resist nuclear weapons effects. *ASCE Manuals & Reports on Engineering Practice, No 42.* American Society of Civil Engineers.

BS8110 (1985) *Structural Use of Concrete.* British Standards Institution.

Balendra, T., Swaddiwudhipong, S, Quek, S., & Lee, S. (1984) Approximate analysis of asymmetric buildings. J. Struct. Div. A.S.C.E, **110**, No. 9, Sept., 2056-2072.

Beck, H. (1962) Contribution to the analysis of coupled shear walls. *Jnl. Amer. Conc. Inst.,* **33**, 1055.

Bhatt, P. (1973) Effect of beam-shearwall junction deformations on the flexibility of connecting beams. *Building Science,* **8**, 149–152.

Biggs, J. M. (1964) Introduction to structural dynamics, McGraw-Hill.

Booth, E. D., Pappin, J. W., & Evans, J. J. (1988) Computer aided analysis methods for the design of earthquake resistant structures. *Proc. Inst. Civil Engrs.,* Part 1, **84**, Aug., 671–691.

Bowles, J. E. (1974) *Analytical and Computer Methods in Foundation Engineering.* McGraw-Hill.

Bowles, J. E. (1982) *Foundation Analysis and Design,* 3rd ed., McGraw-Hill.

Branson, D. E. (1977) *Deformation of Concrete Structures.* McGraw-Hill.

Bulson, P. S. (1985) *Buried structures: static and dynamic strength.* Chapman & Hall.

Cheung, Y. K. & Zienkiewicz, O. C. (1965) Plates and tanks on elastic foundations. *Int. J. Solids Struct.,* **1**, No. 4, 451.

Clark, L. A. (1976) The provision of tension and compression reinforcement to resist in-plane forces. *Magazine of Concrete Research,* **8**, No. 94, Mar., 3–12.

Clough, R. W. & Penzien, J. (1982) *Dynamics of Structures.* McGraw-Hill.

Coates, R. C., Coutie, M. G., & Kong, F. K. (1980) *Structural Analysis.* 2nd Ed., Nelson.

Concrete Society. (1984) Bibliography on concrete structures for hazard protection. *Technical Paper No 107.*

Cook, R. D. (1981) *Concepts and Applications of Finite Element Analysis.* 2nd Ed., Wiley.

Cope, R. J. and Clark, L. A. (1984) *Concrete Slabs — Analysis and Design.* Elsevier Applied Science.

Coull, A. & Choudhury, J. R. (Sept 1967a) Stresses and deflections in coupled shear walls. *Proc. ACI Journal*, **64**, 65–72.

Coull, A. & Choudhury, J. R. (Sept 1967b) Analysis of coupled shear walls. *Proc. J. Amer. Concrete Institute*, **64**, 587–593.

Coull, A. & Wong, Y. C. (March 1981) Bending stiffness of floor slabs in cross-wall structures. *Proc. Instn. of Civil Engrs.*, **71**, Part 2, 17–35.

Coull, A. & Wong, Y. C. (March 1984) Composite action between slabs and lintel beams. *Struct. Enging*, ASCE, **110**, No. 3, 575–588.

Cowper, G. R. (June 1956) The shear coefficient in Timoshenko's beam theory. *J. App. Mechs*, pp. 333–340.

Davies, J. M. & Bryan, E. R. (1982) *Manual of Stressed Skin Diaphragm Design.* Granada.

Dawe, D. J. (1984) *Matrix and Finite Element Displacement Analysis of Structures.* Clarendon Press, Oxford.

Department of the Army, the Navy and the Air Force. (1969) *Technical Manual No. 5-1300. Navy Publication NAVPAC P-378*, Washington, D.C.

Ellis, B. R. (Sept. 1980) An assessment of the accuracy of predicting the fundamental natural frequencies of buildings and the implications concerning dynamic analysis. *Proc. Instn. Civil Engrs.*, **69**, Part 2, pp. 763–766.

Fernando, J. S. & Kemp, K. O. (Mar. 1978) A generalised strip deflection method of reinforced concrete slab design. *Proc. Instn. Civil Engrs*, **65**, Part 2, pp. 163–174.

Gallacher, R. H. (1975) *Finite Element Analysis Fundamentals.* Prentice-Hall.

Gilbert, R. I. (1988) *Time Effects in Conctete Structures.* Elsevier Applied Science.

Glockner, P. G. (1973) Symmetry in structural mechanics. *J. Struct. Div.* ASCE, **99**, ST1, Jan., 71–90.

Green, D. R. (1972) The interaction of solid shear walls and their supporting structure. Building Science, **7**, 239–248.

Green, N. B. (1987) *Earthquake Resistant Design and Construction.* 3rd Edition, Elsevier.

Hambly, E. C. (1976) *Bridge Deck Behaviour.* Chapman & Hall.

Heidebrecht, A. C. & Swift, R. D. (May 1971) Analysis of asymmetrical coupled shear walls. *J. Struct. Div.* ASCE, **97**, No ST5, 1407–1422.

Heidebrecht, A. C. & Smith, B. S. (Feb. 1973) Approximate analysis of tall wall-frame structures. *J. Struct. Div.* ASCE, **99**, No. ST2, 199–221.

Hemsley, J. A. (1988) Comparative flexure of an infinite plate in Winkler springs and a half space. *Proc. Inst. Civil Engrs.*, **85**, Part 2, Dec., 665–687.

Hetenyi, M. (1946) *Beams on Elastic Foundations.* Univ. of Michigan, Series Vol. XVI.

Hooper, J. A. & Wood, L. A. (1976) Foundation analysis of a cross wall structure.

In: Green, D. R. & MacLeod, I. A. (eds.) *Performance of Building Structures.* Pentech Press, pp. 229–268.

Hooper, J. A. (Sept. 1983) Interactive analysis of foundations on horizontally variable strata. *Proc. Inst. of Civil Engrs.*, **75**, Part 2, pp. 475–490.

Hooper, J. A. (August 1984) Raft analysis and design — some practical examples. *The Structural Engineer*, **62A**, No. 8, 233–244.

Horne, M. R. (June 1975) An approximate method for calculating the elastic critical loads of multi-storey plane frames. *The Structural Engineer*, **53**, No. 6, Jun., 242–248.

Horne, M. R. & Morris, L. J. (1981) *Plastic Design of Low-rise Frames.* Collins.

Hrabok, M. M. & Hrudey, T. M. (April 1983) Finite element analysis in design of floor systems. *J. Struct. Engng., Struct. Divn. ASCE*, **109**, No. 4, 909–925.

Hrennikoff, A. (1941) Solution of problems of elasticity by the framework method. *J. App. Mechs.*, **8**, A146.

Institution of Structural Engineers (1989) *Structure soil interaction — A state of the art report.*

Irvine, M. H. (1986) *Structural Dynamics for the Practising Engineer.* Allen & Unwin.

Irons, B. & Ahmad. S. (1980) *Techniques of finite elements.* Ellis Horwood.

Irons, B. M. (1976) The semiloof shell element. In: Ashwell, D. G. and Gallagher, R. H. (Eds). *Finite Elements for Their Shells and Curved Membranes.* Wiley, pp. 197–222.

Irwin, A. W. (1984) Design of shear wall buildings. *CIRIA Report No. 102.*

Kinney, G. F. & Graham, K. J. (1985) *Explosive Shocks in Air.* 2nd ed. Springer-Verlag, New York

Kowalski, R. (1979) *Logic for Problem Solving.* North-Holland.

Livesley, R. K. (1975) *Matrix Methods of Structural Analysis.* 2nd ed. Pergamon Press.

LUSAS: *Finite Element Stress Analysis System.* (1987) Finite Element Analysis Ltd., London.

McHenry, D. (1943) A lattice analogy for the solution of plane stress problems. *J. Instn. of Civil Engrs.*, **21**, No. 2.

MacLeod, I. A. (Sept. 1973) Analysis of shear wall buildings by the frame method. *Proc. Instn. of Civil Engrs.*, **55**, 593–603.

MacLeod, I. A. (Dec. 1976) General frame element for shear wall analysis. *Proc. Instn. of Civil Engrs.*, Part 2, **61**, 785–790.

MacLeod, I. A. (1967) Lateral stiffness of shear walls with openings. In Coull, A. & Stafford Smith, B. (eds.) *Tall Buildings*, pp. 223–244.

MacLeod, I. A. (1971) *Shear Wall-frame Interaction — a design aid with commentary.* Portland Cement Association, U.S.A.

MacLeod, I. A. (1988) Guidelines for checking computer analysis of building structures, construction industry research and information association, London.

MacLeod, I. A. & Abu-El-Magd, S. (1980) Estimation of stiffness of building facade walls undergoing differential settlement. *Int. J. of Masonry Construction*, **1**, No. 2, 46–51.

MacLeod, I. A. & Green, D. R. (Feb. 1973) Frame idealisation for shear wall support systems. *The Structural Engineer*, **51**, No. 2, 71–74.

MacLeod, I. A. & Hosny, H. (Oct. 1977) Frame analysis of shear-wall cores. *J. Struct. Div. ASCE*, **103**, No. ST10, 2037–2047.

MacLeod, I. A. & Marshall, J. (1983) Elastic stability of building structures. In: Morris, L. J. (ed.) *Instability and Plastic Collapse of Steel Structures, Granada*, pp. 75–85.

MacLeod, I. A., Green, D. R., Wilson, W., & Bhatt, P. (1972) Two dimensional treatment of complex structures. *Proc. Instn. of Civil Engrs., Tech. Note* **78**, Part 2, Dec, 589–596.

Marshall, J. (1970) Derivation of torsion formulas for multiply connected thick-walled rectangular sections. *J. App. Mechs*, Paper 70-APM-W.

Melosh, R. J. (1963) Basis of derivation of matrices for the direct stiffness method. *J.A.I.A.A.*, **1**, 1631–1637.

Michael, D. (1967) The effect of local wall deformations on the elastic interaction of cross walls coupled by beams. In: Coull, A. & Stafford Smith, B. (eds.) *Tall Buildings*. Pergamon, pp. 253–272.

Mindlin, R. D. (1951) Influence of rotary inertia and shear on flexural motions of isotropic elastic plates. *J. Applied Mechs.*, **12A**, 69–77.

Meyerhof, G. (1947) The settlement analysis of building frames. *The structural engineer*, Vol. 25, pp. 369–409.

NAFEMS (1986) *A Finite Element Primer*. Dept. of Trade & Industry, National Engineering Laboratory, East Kilbride, Glasgow G75 0QJ, UK.

Paulay, T. (1969) Reinforced concrete shear walls. *New Zealand Engineering (Wellington)*, **24**, No. 10, Oct. 15, 315–321.

Pian, T. H. H. (Nov. 1966) Element stiffness-matrices for boundary compatibility and for prescribed boundary stress. *Proc. 1st Conf. on Matrix Methods in Structural Mechanics*, Dayton, Ohio, Oct 1966, AFFDL-TR-66-80.

Popper, K. R. (1977) *The Logic of Scientific Discovery*. Hutchison.

Poulos, H. G. & Davis, E. H. (1974) *Elastic Solutions for Soil and Rock Mechanics*. Wiley.

Poulos, H. G & Davis, E. H. (1980) *Pile Foundation Analysis and Design*. Wiley.

Quadeer, A. and Stafford Smith, B. (1969) The bending stiffness of slabs connecting shear walls. *ACI Journal*, **66**, No. 6, June, 464–473.

Reilly, R. J. and Funkhouse, D. W. (July 1972) Stiffness analysis of grids including warping. *J. Struct. Div. ASCE*, **89**, No ST7, 1511–1524.

Roark, R J. (1965) *Formulas for Stress and Strain*. McGraw-Hill Kogakusha.

Rockey, K. C., Evans, H. R., Griffiths, D. W., & Nethercot, D. A. (1975) *The Finite Element Method*. Crosby Lockwood.

Rosman, R. (1964) Approximate analysis of shear walls subject to lateral loads. *J. Amer. Conc. Inst.*, **35**, 717.

Ross, C. T. F. (1985) *Finite Element Methods in Structural Mechanics*. Ellis Horwood.

Smith, B. S. & Taranath, B. (Sept. 1972) The analysis of tall core supported structures. *Proc. Instn. Civil Engrs.*, **53**, Part 2, 173–188.

Smith, I. M. (1982) *Programming the Finite Element Method*. Wiley.

Stafford Smith, B. & Riddinigton, J. R. (March 1978) The design of masonry infilled steel frames for bracing structures. *The Structural Engineer*, **56B**, No. 1, 1–7.

Stevens, L. K. (1967) Elastic stability of practical multi-storey frames. *Proc. Instn. of Civil Engrs.*, **36**, 99.

Tamberg, K. G. & Mikluchin P. T. (1973) Torsional phenomena analysis and concrete structure design. *American Concrete Institute*, SP-35, 1–102.

Timoshenko, S. P. & Woinowsky-Krieger, S. (1959) *Theory of Plates and Shells*. 2nd ed., McGraw-Hill Kogakushua.

Troitsky, M. S. (1976) *Orthotropic Bridges: theory and design*. The James P. Lincoln Arc Welding Foundation, Cleveland, Ohio.

Vlasov, V. Z. (1961) *Thin Walled Elastic Beams*. Israeli program for scientific translations.

Walder, V. & Green, D. (1981) The finite element programs FLASH 2 and STATIK. *A Handbook of Finite Element Systems*, CML Publications, Southampton.

Whittle, R. T. (1985) Design of reinforced concrete flat slabs to BS8110. *CIRIA Report 110*.

Wood, L. A. (1977) The economic analysis of raft foundations. *Int. J. Numerical & Analytical Methods in Geomechanics*, **1**, Wiley.

Wood, R. H. (Feb 1969) The reinforcement of slabs in accordance with a predetermined field of moments. *Concrete*, **2**, No. 2, 69–76.

Wyatt, T. A. (1989) Design guide on the vibration of floors. *SCI Publication 076*, Steel Construction Institute/Construction Research and Information Association.

Zbirohowski-Koscia, K. (1967) *Thin Walled Beams*. Crosby Lockwood.

Zienkiewicz, O. C. (1977) *The Finite Element Method*. 3rd ed., McGraw-Hill.

Index